Sport Nutrition

The Carbohydrate Loading Regimen

for

Football

by

Stephen John Shroyer M.D.

Team Physician

In Memorial

1995 Neptune High School Football Championship

Clark, Hailman, Mayo, Hubbard, and Weedon

Hailman, Amabile, and Cella

Game Day in the Meadowlands

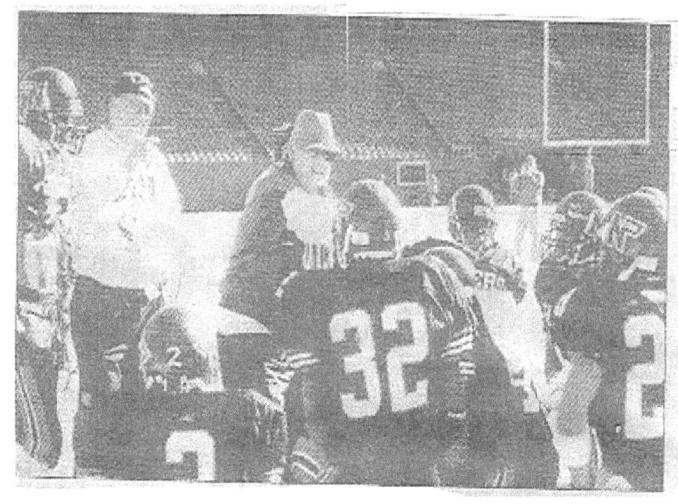

Preface

In the early 90s, I was hired to be the school physician for the Neptune School System. Part of my duties was to attend the home football games. This is where I met and became friends with the Head Coach, John Amabile. After 1 season of play, I discussed with the coach the need for a Nutrition Program to counter the fatigue during the latter half of the football game. John had been coaching Neptune for about 10 years prior to my entrance onto the scene. He was on the verge of winning a Championship for Neptune; the last being 30 years prior. So I went out and developed an Athletic Nutrition Program for the Varsity Football Team. I engineered the program with the following attributes: Carbohydrate-loading meals: dinner and breakfast and pregame hydration with vitamins and ginseng. The in house food company ran the program in the school. The Head coach made it mandatory to attend. With the program up and running, the feedback from the coaches and players were very positive. The meals and the hydration were well tolerated. Neptune went on to win 3 State Championships starting in 1995. The game was played in Giant Stadium, N.J. against the great Franklin High School. Neptune played flawless under the QB, Justin Cella. Coach Amabile became the greatest head coach of Neptune Football in the history of the school.

This book will cover the medical, engineering aspects, to the solid nutrition portion of this program. Nutrition provides the energy to play the game of Football. The nutrients,Carbohydrates, Lipids and Proteins, are studied. Chapter 1, Introduction, gives a description of the program. Chapters 2-4 cover the Carbohydrates. Chapter 5 & 6 cover Lipids and Proteins respectively. Chapters 7-9 cover the energy side of metabolism. Chapters 10 & 11 cover the concept of biological combustion. Chapter 12 is a summary chapter of chapters 7-11. The book ends with chapter 13 which describes the use of Vitamins, Ginseng and Royal Jelly.

The core of the book is dedicated to the organics of biochemistry. It is a treasure of both theoretical and known reaction mechanisms. These reaction mechanisms portray the beauty that life does act in accordance to physical and chemical laws. The scientist at heart will find great pleasure in these sections. Many have wondered how these reactions take place. Here is the pot of gold. The lay person may find it difficult to follow. I would recommend the summary chapter.

This book is intended for para-medical and medical professionals interested in an Athletic Nutrition Program for Football which includes Hydration, Carbohydrate-loading, Vitamins, Ginseng and Royal Jelly.

I would like to thank my family for their patience in this endeavor and to the great coach, John Amabile, for allowing me to be myself and for the opportunity gained in the field of sports medicine.

Table of Contents

Chapter 1

Introduction

General Background

Carbohydrate Loading is an attempt to increase the glycogen storage in the muscles by increasing the intake of carbohydrates. Not only is it important to increase the muscle content, but also the liver. The liver is usually depleted of its glycogen stores in the over-night fasting of sleep. This maintains blood sugar during the fasting period. Glycogen is a polymer of glucose which is used to store energy for **Catabolism** (the breakdown of compounds to produce energy for the use of work). The work, in our case, is the successful playing of a game of football (exercise). The energy could also be used for the creation of compounds (synthesis of proteins), **Anabolism**. In the plant kingdom, the glycogen is replace by starch. Starch is similar in structure to glycogen. It serves the same purpose. We can digest and absorb starch and therefore utilize it's energy potential. A common form of starch is the potato and rice. These foods do not alter blood sugar much because of the way we need to digest and absorb the starch. So they are blood sugar friendly.

Metabolism

Anabolism	Catabolism
Synthesis (building-requires energy)	Destruction (breakdown-release energy)

Glucose polymer

Animal	Plant
Glycogen	Starch

Polymer is a macro-molecule made from single units called monomers, a glucose molecule, linked together forming a large chain. As we will see, these are highly branched chains and have different linkages holding the glucose molecules together. A linkage is a covalent bond.

Catabolism is a form of combustion. But not like the burning things. If we burned glucose, we would end up with Carbon dioxide, Water, and **heat** (energy). Obviously this is done on a bench-top in an organic chemistry lab. In living systems, the breakdown of glucose through the oxidation-reduction reactions create energy-rich compounds, mainly Adenosine Tri-Phosphate. The breaking of the phosphate bond, which is very high in energy, releases the energy, known as free energy, to do work. ATP is a form of potential energy, usable energy. Heat is also a by-product which is transferred to water and dissipated by sweating. Water modulates the heating process, but when it reaches high levels, sweating occurs. Sweating assists through evaporation.

Catabolism

Non-living system	Living system
Combustion	**Oxidation-reduction**
Reactants: Heat + excess oxygen	Glucose + enzyme (re-usable)
Products: CO_2, Water, and Heat	CO_2, H_2O, free-energy $\rightarrow$ ATP
$\rightarrow$ environment	ATP $\rightarrow$ work + heat
	heat $\rightarrow$ water $\rightarrow$ sweat
Process: With heat as energy, Oxygen is added to C and H atoms.	With Electron transfer, Oxygen is added to C and H atoms. O_2 is the ultimate electron acceptor O_2 can covalent bond with H + C.

Oxidation-reduction (Redox) reaction

Oxidation	Reduction
reductant : electron lost (- (-1) = +1)	oxidant : electron gained (+ (-1) = -1)
oxidation # increasing (more (+))	oxidation # decreasing (more (-))
Oxidizing agent is reduced	Reducing agent is oxidized.

Formal Q (charge) is the net oxidation # assigned to an atom.

Example:　　$H_2 + F_2 \rightarrow 2\ HF$

$H - H\ \ +\ \ F - F\ \ \rightarrow 2\ H - F$　　$-$ = (covalent bond) = 2 e
　　　　　　　　　　　　　　　　　　　2 e with no bond = **lone pair e**

$H_2\ (0) \rightarrow 2\ H\ (+) + 2\ e$　　　$0 \rightarrow +1$　H is oxidized.

$F_2\ (0) + 2\ e \rightarrow 2\ F\ (-)$　　　$0 \rightarrow -1$　F is reduced.

　　　$H - F$ is a covalent, polarized bond. F is more electronegative than H. The F is partially negative charged and H is partially positive charged.
　　　F (-) has 8, 1 more, electrons in it's valence shell; normally 7 is present.
　　　H (+) has 0, 1 electron absent, in it's valence shell; normally 1 is present.
　　　These are ions, charged atoms and highly reactive.

The Carbohydrate loading is mainly used for marathon runner, swimmer, or cyclist. It is usually a 90 minute high-intensity competition. Loading is less necessary for shorter activities.

In High School Varsity Football, the event is broken down as follows:

1. Pregame meeting and preparation.
2. 2 quarters of play at 12 minutes each.
3. Half-time of 20 minutes or so.
4. 2 quarters of play at 12 minutes each.
5. Post game meeting and preparation.

Total time could very well be 6-8 hours of the day. At least 4 hours is in full uniform. There is a lot of heat generated along the way making energy necessary to dissipate the heat. This is an energy driven process requiring a lot of calories.

The process is approximately a 7 day one. There is a depletion type, non-depletion type, and a short workout type.

In the depletion type, on the 7th day prior, you would perform an exhaustive exercise to deplete the glycogen stores. For the next 3 days (6,5,4), you would do low exercise only. During days 7 to 4, you would reduce or maintain your carbohydrate intake at about 50-55% of your total calories. You would increase protein and fat to compensate the carbohydrate deficit. Then the next 3 days (3,2,1), you would eat a high carbohydrate diet, an increase to 70% of your daily calories or about **4.5 g carbohydrates per lb**. of body weight. Now you would cut back on protein and fat because of the high carbohydrate diet. During day 3, you would continue the low exercise program. Day 2 would be a day of rest. This would prevent depleting the carbohydrate loading regimen. This was thought to store more glycogen because of the starvation period.

The **non-depletion** type, there was no starvation. You only increased your carbohydrate diet to 70% and decreased your training 3 days prior to the event. This is the regimen we did at the high school football level.

The short workout calls for a normal diet with light training until the day before the event. On the day before the event, you perform a very short, extremely high-intensity workout then consume 12 g carbohydrate per kg of lean mass over the next 24 hours.

At Neptune High School, our week went as follows:
1. Game every Saturday at 2 PM.
2. Practice Monday-Friday.
    1. Monday to Wednesday a hard workout.
    2. Thursday a lighter workout.
    3. Friday a very light workout.
3. Players were eating regular meals during the week.
    1. Friday dinner was 4-600 calorie (**70%** carbohydrate, 15% fat and 15% protein) loading meal in the cafeteria after practice.
The Athletic Nutrition Program was **mandatory** for all players.
    2. Saturday breakfast was 4-600 calorie (70% carbohydrate, 15% fat, and 15% protein) loading meal in the cafeteria before game meeting and preparation.

3. Between 11-12 noon, the players hydrated with **Cranapple** fruit drink, 4 Liters. They also took 1 **Centrum** vitamin and 1 High-energy **ginseng** supplement (tablet form). Also a snack bar, such as granola, would have been ideal, but was not done.
4. Players emptied their bladders 30 minutes prior to game time.
5. During the game, water was administered.
6. Post-game hydration was left to the players.

The program was well tolerated by the players. There was no complications from the program.

The program at Neptune was conducted at 1/3 of the regimen both in the hydration and nutrition due to budget restraints and school board approval. The full program would be 3 days duration. This would have been ideal.

This upcoming chapters will proceed to describe in detail the nutrition regimen of the program including vitamins and ginseng use. The hydration part was fully discussed in book 1, **Sport Hydration: A Synopsis on Concepts and Applications** which you can find and purchase at www.Lulu.com.

Carbohydrates are found in **grains**, **dairy** products, **fruits**, **vegetables** and **legumes** (beans and peas) They are also in **sugar** and **sweets**.

The living systems have a priority in the use of nutrients for energy. They prefer carbohydrates first, fats second, and proteins last. If there is excess carbohydrates, they store them first as glycogen , then as fat. Proteins are used to build protein. But in extreme starvation, proteins well be used as energy.

The brain's main energy catabolite is glucose. In starvation, it can use fat, carbohydrate, protein derivatives.

The glycogen in the liver controls blood sugar levels mainly for brain metabolism.

The glycogen in the muscle is used for muscle exertion.

The dinner in the evening restores the muscles, while the breakfast restores the liver. After sleeping all night, the liver is depleted of its glycogen stores, but the muscles are loaded from resting.

After a 4-600 calorie meal, approximately 4 hours is needed for complete digestion. If there is only about 2 hours prior to game time, you can consume a snack of about 200 calories for energy.

Example of Breakfast

Item (amount)	Carbohydrate (grams)	Calories
Milk, fat-free (12 ounces)	18	29
Nutty barley cereal (1 cup = 8 oz.)	92	413
1 cinnamon raisin bagel (3.5 inch diameter)	39	194
Reduced-calorie margarine (1 Tbs.)	0	51

This is an example of a breakfast for a marathon runner. To modify this for our football program, we would decrease the cereal by 2 oz. to give us a total calorie count closer to 600. The cereal calories would go from 413 to 313.

Some conversions to remember: @ gm of carbohydrate or protein = 4 calories.
@ gm of fat = 9 calories.

Let us analyze the breakfast.

Milk: 18 * 4= 72 calories; 129-72 = 57 calories of fat/protein; 72/129 = 55.81 % Carbohydrate.

Cereal: 92 * 4 = 368; 413 – 368 = 45; 368/413 = 89.10% (0.8910 *100 = %)

Bagel: 39*4 = 156; 194-156 = 38; 156/194 = 80.41%

Margarine: 0*4 = 0; 51-0 = 51; 0/51 = 0% All 51 calories from fat mainly.

Overall carbohydrate loading calories: 55.81 + 89.10 + 80.41 = 225.32 /3=75.1067 %
Total calories: 129+413+194+51 = 787 calories.

Can you calculate the % and total calories if the cereal amount is reduced by 2 oz.?

Hint: There are 8 oz. in 1 cup. 2/8 = ¼ = 25%.

Answer: 74.8%; 687 calories.

Example of Dinner

Item (amount)	Carbohydrate (grams)	Calories
Salmon, baked (3 oz.)	0	175
Brown rice (1.5 c.)	67	324
Broccoli, steamed (1 c.)	11	54
Milk, fat free (8 oz.)	12	86
Lettuce salad with tomatoes and carrots (1.25 c.)	3	16
Fat free Italian salad dressing (2 Tbs.)	2	14

Walnuts (0.25 c.)   4   191

3 of the highest sources of carbohydrates are rice, milk, and broccoli. Rice has starch, a complex carbohydrate.
2 of the highest sources of fat are Salmon and Walnuts. Salmon has the omega-3 essential fatty acids.

Analysis:

Salmon: 0*4 = 0; 175-0 = 175; 0/175 = 0% All calories from fat and protein.

Rice: 67*4 = 268; 324-268 = 56; 268/324 = 82.72%

Broccoli: 11*4 = 44; 54-44 = 10; 81.48%

Milk: 12*4 = 48; 86-48 = 38; 48/86 = 55.81%

Salad: 3*4 = 12; 16-12 = 4; 12/16 = 75.0%

Salad dressing: 2*4 = 8; 14-8 = 5; 8/14 = 57.14%

Walnuts: 4*4 = 16; 191-16= 175; 16/191 = 8.37%

Overall carbohydrate % = 360.52/5 = 72.104%
Total calories = 860 calories

How would you modify this meal to meet the goal of 600 calories?

Hint: Get rid of fat.

Answer: Eliminate Walnuts. 72.104%; 669 calories.

Some more conversions:
        1 c. = 8 oz.
        2 c. = 1 Qt.
        4 Qt. = 1 Gal.
        1 Qt. Approximately = 1 L
        16 oz. = 1 Lb.
        1 oz. = 1 Tbs.= 30 grams
        3 tsp = 1 Tbs.

Now, we will do a different kind of analysis. We go to the pantry and pull out 2 high carbohydrate items and prepare them for breakfast and dinner.
The first item is pancakes. The Nutrition Facts label states the following:

Serving Size: 1/3 c. = 3 4" (diameter) pancakes
Calories 150
Total Fat 1.5 g.
Cholesterol < 5 mg
Total Carbohydrate 31 g.
Protein 4 g.

To make 400 calories, how many pancakes? Can you approximate the calories from the label constituents?

400/150=2.67; 2.67*3=8 pancakes.

31*4=124; 4*4=16;1.5*9=13.5 124+16+13.5=153.9 calories.

Second item is white rice. The label states the following:

Serving size: ¼ c.
Calories 150
Total Fat 0
Total Carbohydrate 35 g.
Total Protein 3 g.

To make 350 calories, how much rice to cook? Calories?

350/150=2.3; 2.3 * 0.25 = 0.583 c.

35*4=140; 3*4=12; 140+12=152 calories.

> Summary: The % ratio in our Carbohydrate loading meals are: 70/15/15.
> All meals need 4 hours for digestion. Ideally, 3 full days of carbohydrate loading is optimal. A 300 calorie snack bar during hydration time is ideal.
> 1 Gallon of Cranapple each day for 3 days is optimal. On game day, it is consumed 2.5 hours prior to game time with voiding 30 minutes prior to start of game.
> 1 Centrum vitamin and 1 Hi-energy ginseng given at hydration.
> All of this loads the liver and muscle with glycogen, the energy for muscle activity and sweating which controls body temperature.

Chapter 2

Carbohydrates

Digestion

There are 2 processes at work, one is condensation and the other hydrolysis. They are identical, but opposites. In condensation, water is removed and 2 molecules are joined together. In hydrolysis, the opposite is true. So in digestion, condensation is reversed by hydrolysis.

Hydrolysis

$$R''- R' + H2O + \text{digestive enzyme} \rightarrow R''OH + R'H$$

R", R' are monosaccharides.

The carbohydrates found in diet are:
  Sucrose: cane sugar
  lactose: milk
  starches: potato, grains
  others: amylose, glycogen, alcohol, lactic acid, pyruvic acid, pectins, dextrins, and carbohydrate derivatives in meats.
  Cellulose: a form of plant polysaccharide, non-digestible because of no enzyme.

The digestive process starts in the mouth and stomach:

  Mouth
    Parotid gland
    ptyalin (α-amylase)
    starch → **maltose** (disaccharide) + oligosaccharides

  Fundus of Stomach
    Amylase stops eventually

  Small Intestine
    Pancreas
    α-amylase
    duodenum/upper jejunum

Enterocyte at microvilli brush border
            lactase: Lactose → galactose + glucose
            sucrase: Sucrose → fructose + glucose
            maltase, α-dextrinase: Maltose, oligosaccharides → glucose
All these monosaccharides are water soluble because of the hydroxy groups associated with them. They all enter the portal blood which leads to the liver.

Summary

Mouth & Stomach
Maltose + Oligosaccharides
Small Intestine
Monosaccharides

Saccharide comes from the Greek word sakcharon, meaning sugar. Monosaccharides are single polyhydroxy aldehyde or ketone units.

Aldose: R-HC=O

Ketose: R-R'C=O

R = Alkyl group
C=O is the carbonyl function
H=hydrogen atom

The monosaccharides can enter Glycolysis to create ATP to do work; such as muscle contraction, can enter Glycogenesis to be stored as glycogen in the liver, or remain as glucose an become glycogen in the muscle. The liver controls blood sugar for all cell function and needs, especially the brain. It also supplies sugar for storage as glycogen in the muscle. The liver stores are much larger than the muscle stores. The muscle stores are depleted in 1 hour of vigorous exercise. 4 quarters of football equals approximately 48 minutes. This is within 60 minutes by 12 minutes. The liver stores can last between 12-24 hours. There is a lactate acid cycle between the muscles and liver to convert the lactic acid which is being build up during muscle exercise to glucose in the liver. The cycle is called the Cori cycle which tries to maintain glucose or glycogen levels for further muscle activity. It also buffers the blood from acid build up, metabolic acidosis.

Monosaccharides → Glycolysis (Glucose breakdown) → ATP (energy to do work)

            → Glycogenesis (Glycogen synthesis) → Liver glycogen stores

            → Glucose → Glycogenesis (Glycogen synthesis) → Muscle glycogen
stores.

Let's take a closer look at the process of **Hydrolysis** as seen in the way of the organic chemist.

Hydrolysis: Alpha-amylase: Breakdown of Starch to Glucose

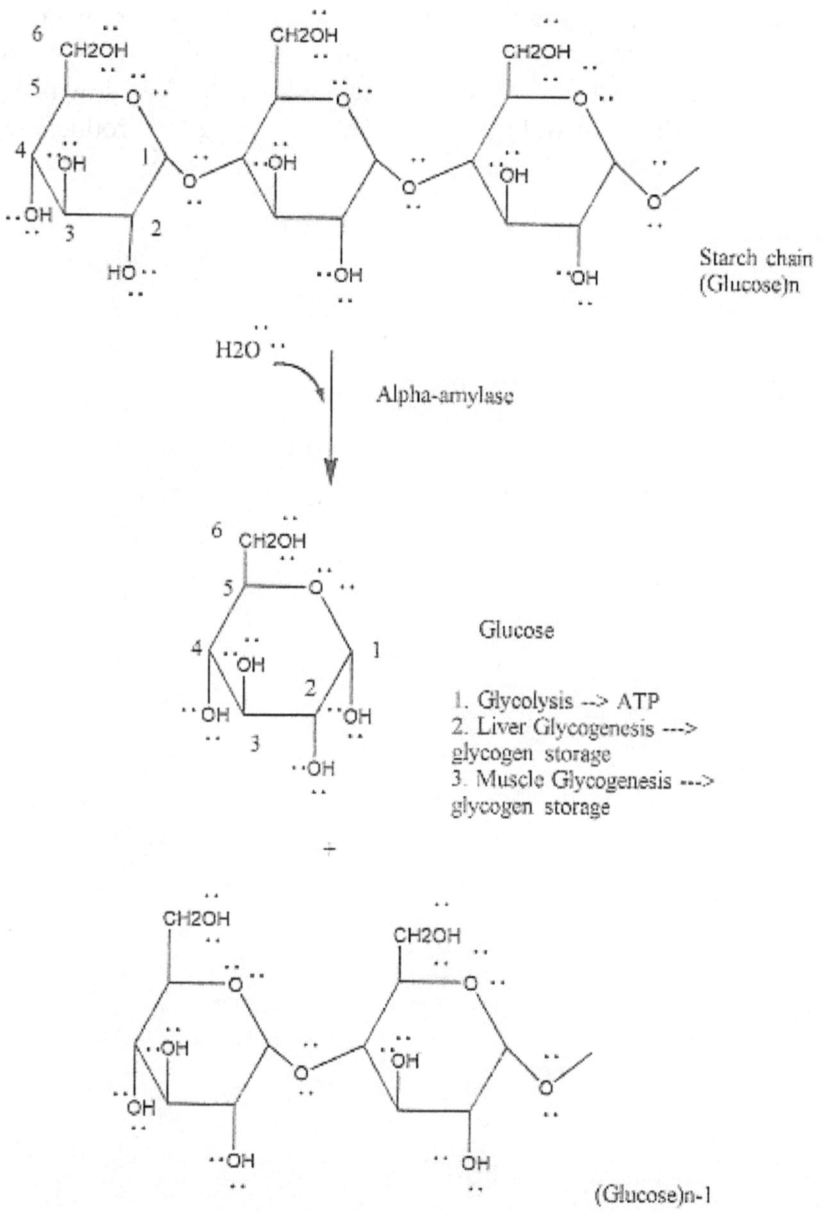

Starch chain
(Glucose)n

$H2O$

Alpha-amylase

Glucose

1. Glycolysis --> ATP
2. Liver Glycogenesis --->
glycogen storage
3. Muscle Glycogenesis --->
glycogen storage

+

(Glucose)n-1

Description of figure: Hydrolysis: Starch to Glucose

Structure: Starch is a glucose (residue=monomer= 1 glucose molecule) polymer = (Glucose)n
                n=# of glucose residues.
                Each glucose is cyclic and portrayed as a Haworth Perspective projection.
                The cyclic ring is an acetal form created by 2 alcohol and aldehyde reactions.
                C1 Oxygen (glycosidic linkage) is in the alpha position which is down.
                Hydrolysis is the addition of water to the molecule making the products water soluble.
The net result is (Glucose)n-1. The starch is broken down 1 residue at a time.

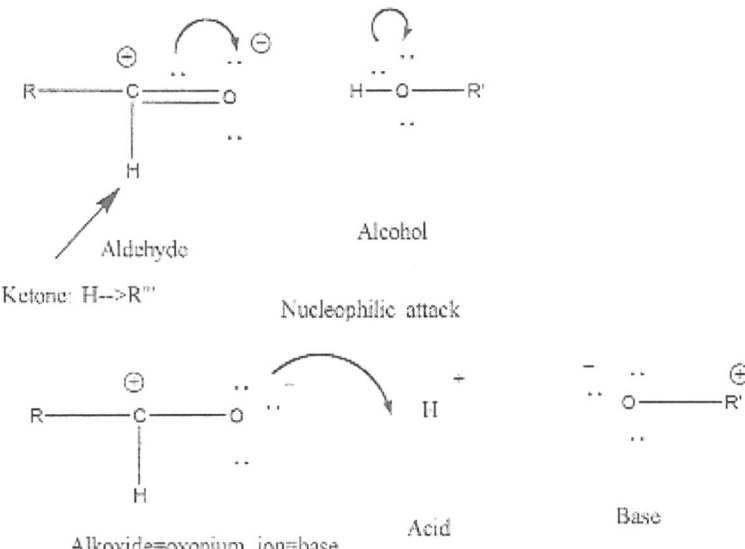

Hemi-acetal/ketal and Acetal/Ketal Formation
Aldehye/Ketone Alcohol reaction: Acid Base reaction

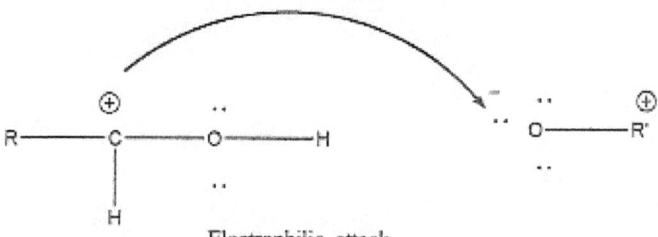

Electrophilic attack

Carbocation:Acid

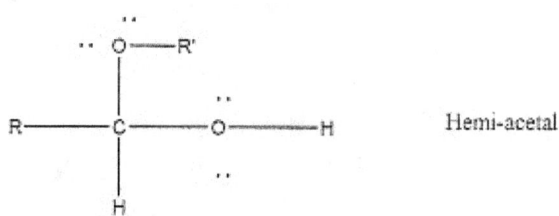

Hemi-acetal

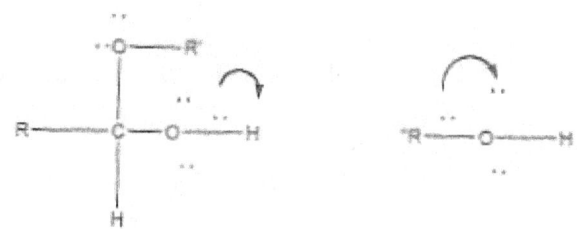

Hemi-acetal=Acid

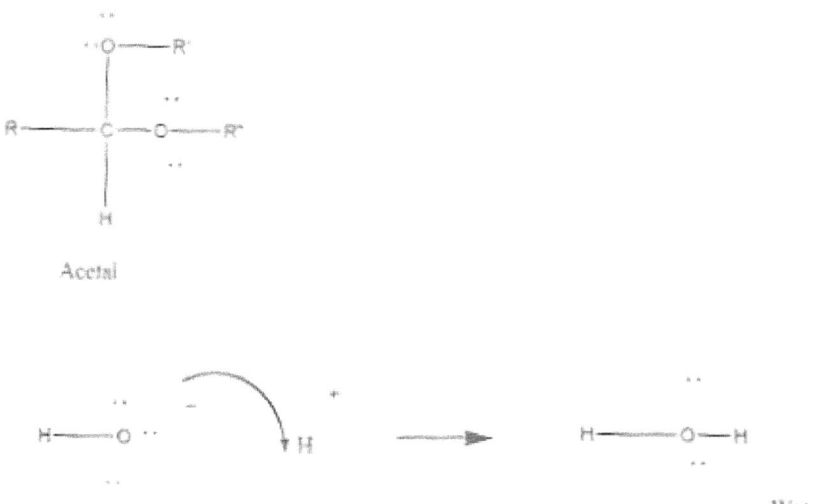

Acetal

Water

The curly arrow represents an electronic change or reaction.
Double Bonds (pi bonds), ions, (lone pair e), and polar sigma bonds (single bonds) are highly reactive.
Circled charges are partial charges (polar) atoms. The carbocation has a formal charge of +1. The partial charge → formal charge. There is equal possibility for either attacks for the reaction mechanism.
Charges are formal charges: integer charges=full charges.
Nucleophilic attack is an attack on vacant electron orbitals. It is not an attack on protons of the nucleus.
Electrophilic attack is an attack on electron filled orbitals.
Alcohol can act as an acid. It is an Alkyl (R group=Hydrocarbon) type of water.
Water itself is an end product.
A covalent bond = 2 e.

Hydrolysis is the addition of water to the glycosidic linkage of a multi-glucose polymer in an acid base reaction.
The glycosidic linkage is similar to a ether molecule: R-O-R. Note the similarity to water: H-O-H.

Self-cyclization of D-Glucose

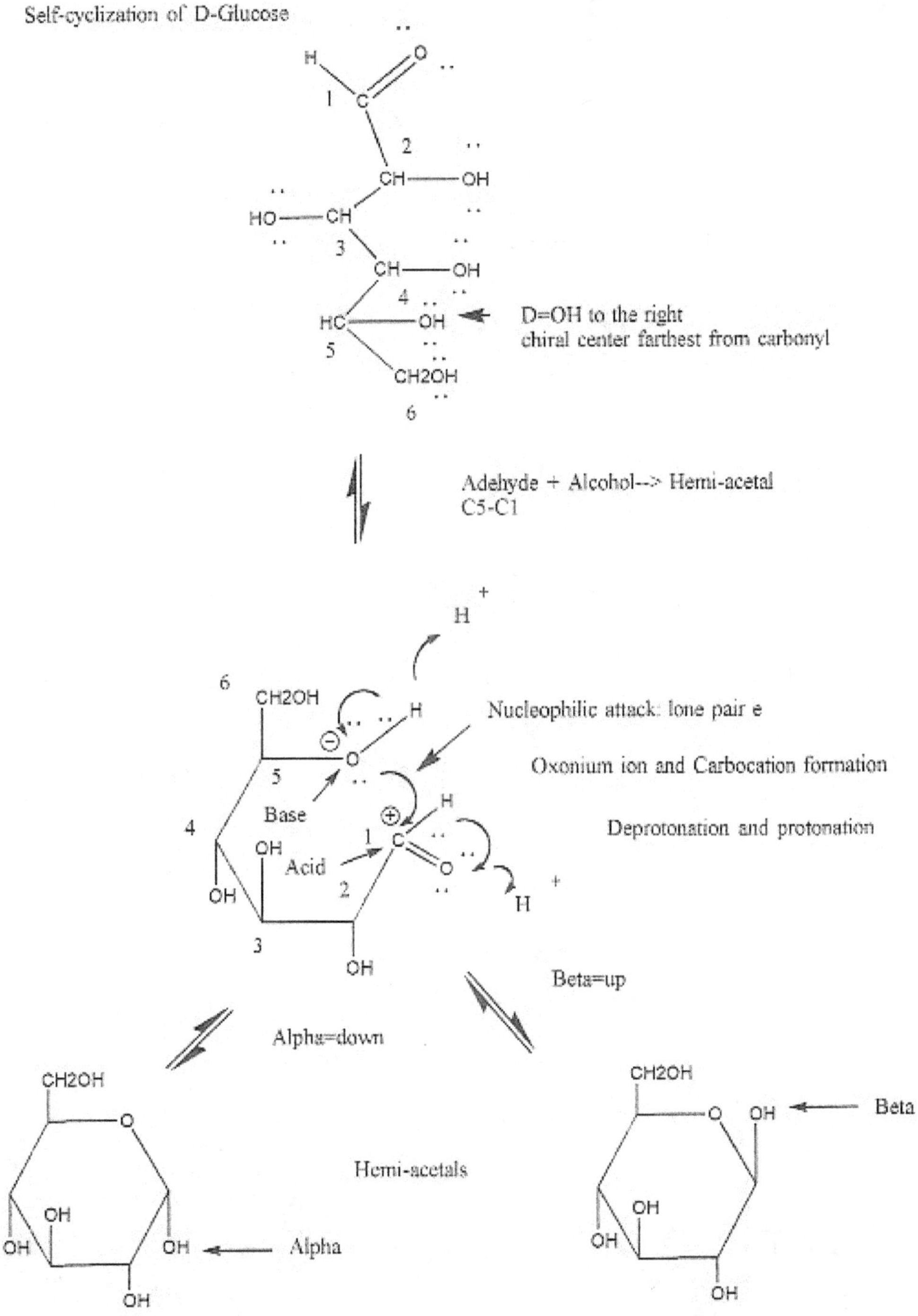

D=OH to the right
chiral center farthest from carbonyl

Adehyde + Alcohol--> Hemi-acetal
C5-C1

Nucleophilic attack: lone pair e

Oxonium ion and Carbocation formation

Deprotonation and protonation

Beta=up

Alpha=down

Hemi-acetals

Alpha

Beta

Base

Acid

Hydrolysis: Reaction mechanism: Acid-Base reaction

Alpha-amylase breaks bond creates oxonium ion = base and carbocation=acid

Carbocation

water   Acid   Base

Nucleophilic attack: lone pair e

Electrophilic attack

Monosaccharide end products of digestion

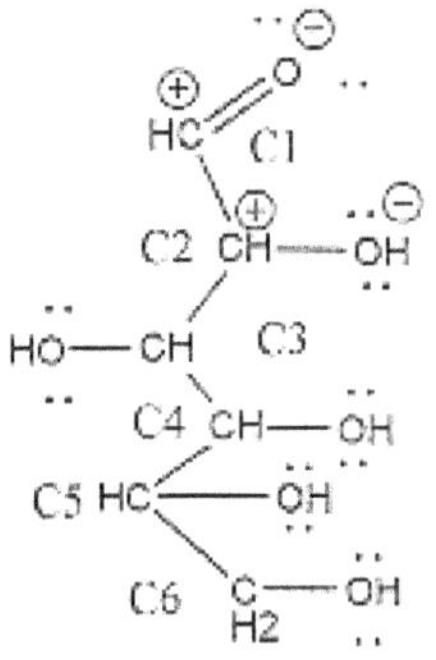

+/- circled: polarized atom covalent bond=polar sigma bond

: lone pair electrons: Lewis Dot Structure

OH =   —O——H

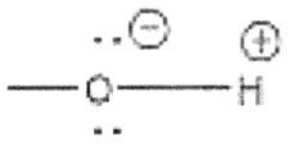

D-OH at C5 to the right

Double Bond: sigma and pi bonds

sigma bond = 2 e
pi bond = 2 e

D-Glucose
Aldose

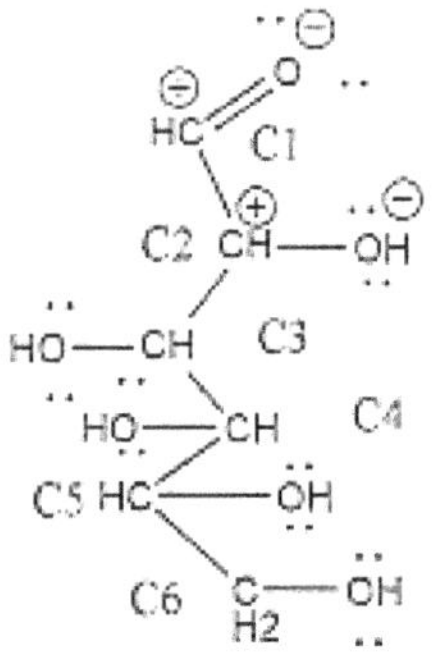

C1: Aldose

D-Galactose
Aldose

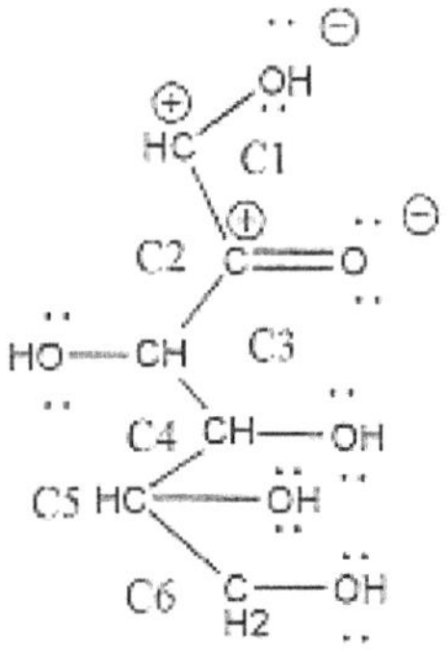

C2: Ketose

D-Fructose
Ketose

Absorption

Of the total carbohydrates absorbed, 80% is glucose and 20% is galactose and fructose. All are done by an Active Transport process.

The Active Transport of sodium is associated with the co-transport of glucose.

1. **A.T.** Na+ ICF → ECF
2. Na+ and glucose **facilitated diffusion** at brush border
   Lumen → ICF (possible because of low ICF Na+ from A.T.)
3. Glucose → paracellular space → blood by **facilitated diffusion.**

Galactose is moved the same as glucose.

Fructose is only **facilitated diffusion**. Fructose in phosphorylated inside the cell and converted to glucose. It's diffusion rate is much slower. Because of the addition of phosphate, the diffusion rate is slower. The phosphate energizes fructose for the conversion process; it targets fructose for conversion; and picks a location for fructose so the process can occur. So it labels, directs, and locates the conversion process for the target: fructose.

Lumen   Blood
142 mEq/L   145 mEq/L

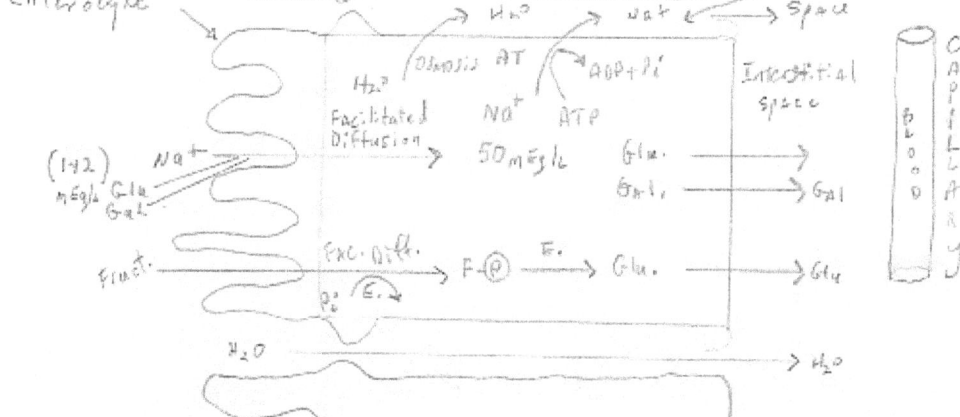

Here we have an **enterocyte**. Note the Na+ concentration of luminal chyme and blood is the same. ICF Na+ concentration is only 42 because of the Active Transport of Na+ against the higher ECF Na+ concentration. The glucose and galactose is co-transported with Na+ and then by facilitated transport to the blood. Fructose enters by facilitated transport but is phosphorylated for conversion to glucose which enters the blood as usual. H2O follows by the force of osmosis. So glucose and galactose is assisted by Na+. Fructose is converted because glucose is preferred. Facilitated diffusion is protein mediated down a concentration gradient. Active Transport is protein mediated with the use of energy against a higher concentration gradient. Osmosis is from a high concentration of water to a lower concentration of water.

Chapter 3

Glycogen

Synthesis

Glycogen synthesis takes place in the liver and skeletal muscle. The starting point is glucose 6-phosphate.

$$\text{D-Glucose (digested)} + ATP \rightarrow \text{D-\textbf{glucose 6-phosphate}} + ADP$$

Digested glucose is phosphorylated (energized) by ATP (a high energy molecule).
This is done by the isozymes **hexokinase** I and II in muscle, hexokinase IV (glucokinase) in liver.
Isozymes are very closely related protein enzymes which are isomers. They are the same molecule, just different in structure.
Hexo = glucose + kinase = phosphorylating enzyme.
Note I & II are in the muscle, while IV is in the liver (compartmentalization).

D-glucose → RBC (Red Blood Cell) → lactate (glycolytically = broken down) → liver → G-6-P by **gluconeogenesis** (synthesis of glucose from on carbohydrate derived substrates). This is called the **Cori Cycle**.

There are 2 requirements needed before glycogen synthesis can take place:
1. **UDP-Glucose** (sugar nucleotide) = pathway first reaction step.
2. **Glycogenin** = glycogen chain initiator.

UDP-Glucose

Sugar nucleotides are used in the transformation or polymerization processes of hexoses. They act as substrates in polymerization or intermediates in the production of derivatives of hexoses. In the synthesis of glycogen, UDP-Glucose donates glucose for Glycogen synthesis. What are it's properties?

1. The condensation reaction which creates a sugar nucleotide has a very small free-energy change. But one of the products of the reaction, PPi, is hyrolyzed and releases -19.2 kJ/mol free-energy change which is strongly exergonic. This makes the reaction and the pathway irreversible; ensuring the forward reaction to glycogen. This is common in polymerization reactions. The reaction becomes favorable in a thermodynamic viewpoint.

NDP-sugar pyrophosphorylase

Line 1 : 2 reactants are sugar phosphate (sugar-P): Alpha-D-Glucose 1-Phosphate (G-1-P) Alpha: carbonyl O is oriented down, carbonyl C is C #1.
D=Dextro=right: The Oxygen of the hydroxyl group on the most distant chiral C atom from the carbonyl group is oriented to the right. In G-1-P, the C atom is #5.
The other reactant is a Nucleoside Triphosphate (NTP) (P-P-P-Ribose-Base): Uridine Triphosphate = 3 Phosphates + Ribose (sugar) + Uridine (base).
The enzyme (catalyst) is _NDP-sugar pyrophosphorylase_: 1. This enzyme places stress on the alpha phosphorous atom in the nucleoside by a stretching process.   A lone pair of electrons jump to the adjacent Oxygen atom creating an electrophilic phosphorous atom at alpha (P+). An electrophile loves electrons and is very reactive. The 2 products are Pyrophosphate (PPi) and Uridine Monophosphate (the electrophile). 2. The G-1-P is de-prontonated at the hydroxyl group on the phosphate group creating a nucleophile. A nucleophile loves a nucleus of protons (this is more an empty electron orbital being more positive in charge). This is an acid base reaction in a water environment.

$$H20 \rightarrow + H \ + \ - OH$$
water   Acid   Base

$$HPO4(-1) + \ -OH \ \rightarrow PO4(-2) + H2O$$
acid   base   base   acid

In part 2 of the reaction, Line 2, the enzyme approximates (closes the distance) between to the 2 reactants making possible an nucleophilic attack to occur between the basic Oxygen on the sugar phosphate and the acidic Phosphorous atom, the electrophile, on the Uridine Monophosphate. The products are PPi and NDP-sugar (Sugar nucleotide) (NDP=Nucleotide DiPhosphate). NDP-sugar in this case is Uridine DiPhosphate.

Lines 3 and 4, PPi is hydrolyzed by the enzyme, inorganic *pyrophosphatase*, releasing a large negative free-energy ensuring the forward reaction; making the reaction irreversible. The enzyme adds water to each phosphate group of PPi. The products are 2 Pi, both inorganic phosphate ions. These are both nucleophilic attacks on the H+ in the aqueous environment. The enzyme stretches the phosphodiester bond creating to basic species, both very reactive, and nucleophiles. These attack the H+ from water dissociation.

The net reaction: Sugar-P + NTP $\rightarrow$ NDP-sugar + 2 Pi.

PPi

Enzyme stress: twist, bend, and stretch

Nucleophilic attack: lone pair electrons

H+

Acid

Base

H+

PO4(-3)

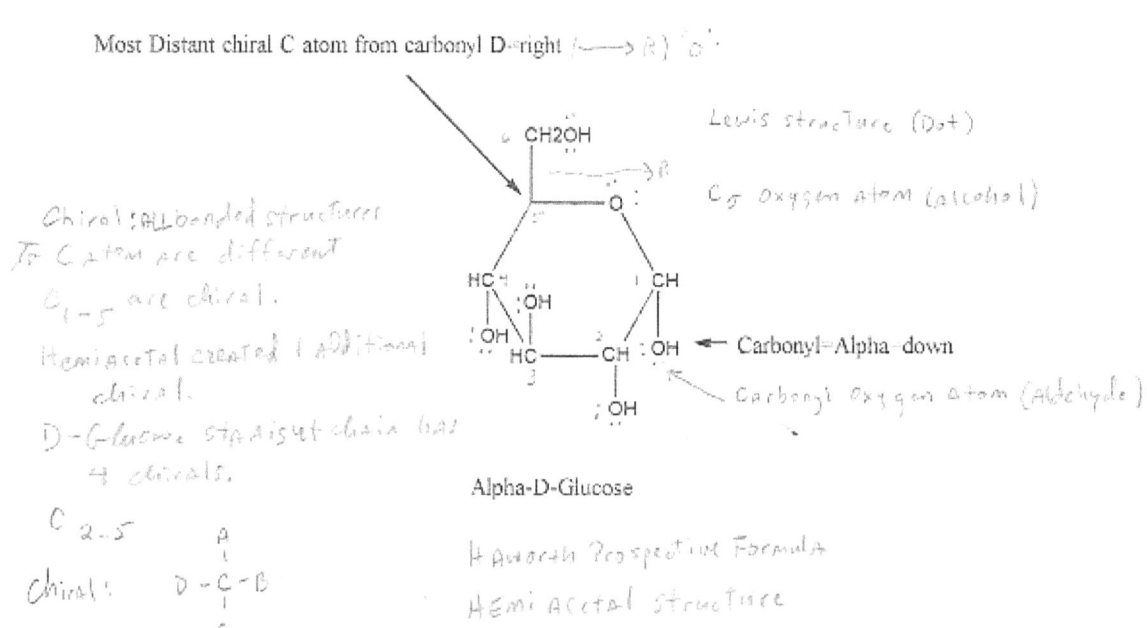

Here is a more detail look at the structure of UDP-Glucose. First, the D-Glucosyl group of UDP-Glucose will be shown.

Most Distant chiral C atom from carbonyl D-right

Alpha-D-Glucose

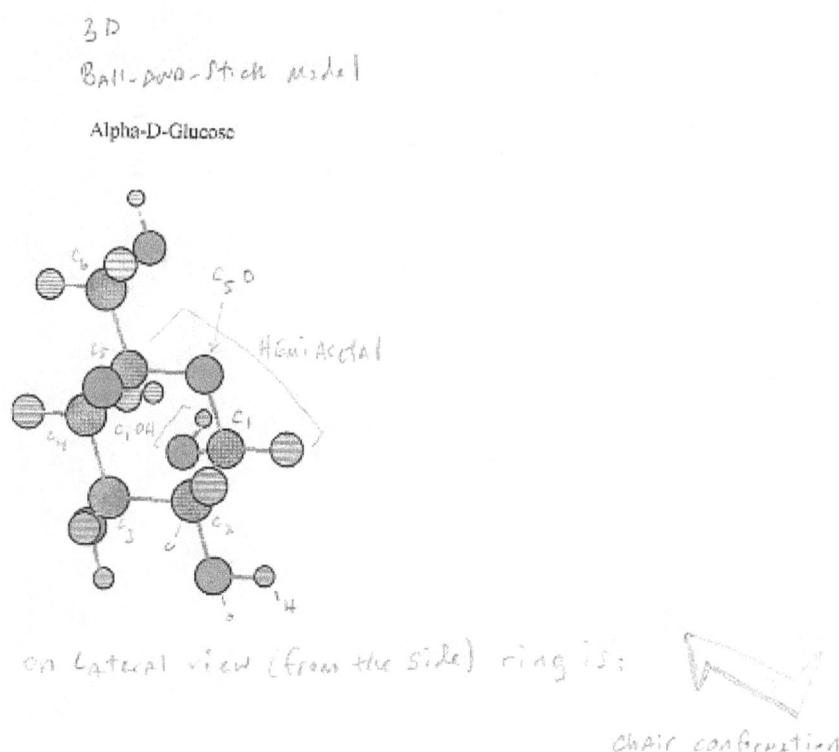

3 D
Ball-and-Stick model

Alpha-D-Glucose

on lateral view (from the side) ring is:

chair configuration

These 2 figures are rather dense in detail; so I will tease out the important aspects of each detail separately.

The open straight chain and cyclic form of glucose is in equilibrium. At times, you will have both forms present; each at a certain %. If we look at the open form, we can make an easy determination as to the D or L isomer present. Historically, this was based on comparing the molecule to D or L -Glyceraldehyde, the smallest monosaccharide – a triose (3C). Look for the next to last C atom in the molecule. Which way is the Hydroxyl group pointing-to the right or left? The projections in the figure are Fischer type. The horizontal lines are towards you and the vertical lines are away from you. In this formula, the Hydroxyl groups facing right will be down in the Haworth perspective formula. Glucose mainly exists in the cyclic form because it is favored thermodynamically. In UDP-Glucose, Glucose becomes a function group, so it is named D-Glucosyl. The e is dropped at the end of the name and yl is added to designate functional group stature.

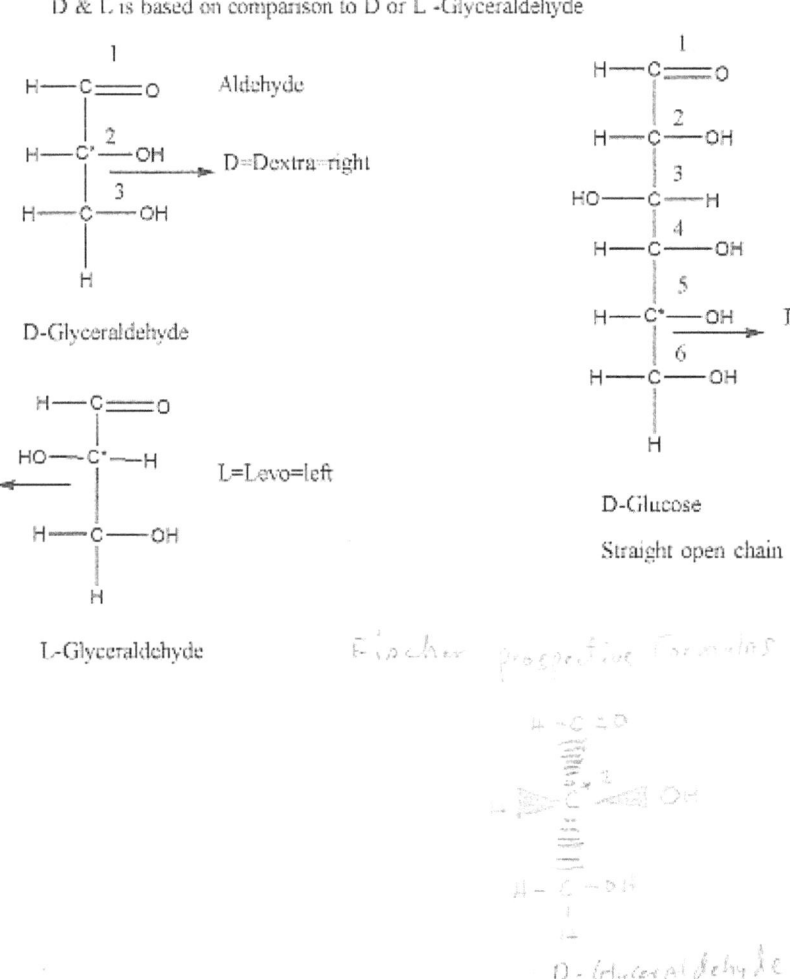

D & L is based on comparison to D or L -Glyceraldehyde

Aldehyde

D=Dextra=right

D-Glyceraldehyde

L=Levo=left

L-Glyceraldehyde

D-Glucose

Straight open chain

In the Haworth perspective formula, the ring structures can take on 2 configurations, the alpha and beta isomers. The C1, the anomeric C atom = the C of the Carbonyl function, will designate the configuration. If the Hydroxyl group in up; it is beta type. If the Hydroxyl group is down; it is alpha type. In the Fischer perspective formula, down is to the right and up is to the left. To be cyclic means that the ring is either a Hemiacetal or Acetal form. In the figure, the H atoms have been removed for clarity. C has 4 covalent bonds. The figure on the right is even more simplified by showing only lines to represent the Hydroxyl group positions.

Alpha & Beta anomers: Haworth perspective formulas
isomers/ configurations

C5 Hydroxy group

6 C—OH

5 C———O

Anomeric C

C1 Carbonyl
group

C 4   Left   1 C

OH

OH   OH

2   OH ← Alpha=down

C———C

3

OH   Right in open chain

Thickened edges facing you

OH ←– Beta=up

HEMIACETAL ;

Conformations are interchangeable. No bonds are broken to do so.

There are 2 conformations in which glucose can take as form. Because of steric hindrance, the carbon backbone chain whether open or cyclic will take the orientation of a staggered tooth form. These forms take on 2 types, the boat or chair. These are the 2 conformation types. Steric hindrance is caused by the repulsion of the electron clouds of each atom in the molecule. Remember like charges repel.

$$(-) \leftarrow \rightarrow (-).$$

Conformations are interchangeable. No bonds are broken to do so.

The functional groups take positions along an axis. The axial groups are parallel to this axis while the equatorial groups are not.

The figure also shows pyran ring in which glucose is considered a pyranose by comparison. You must compare using the Haworth perspective formula. Try it.

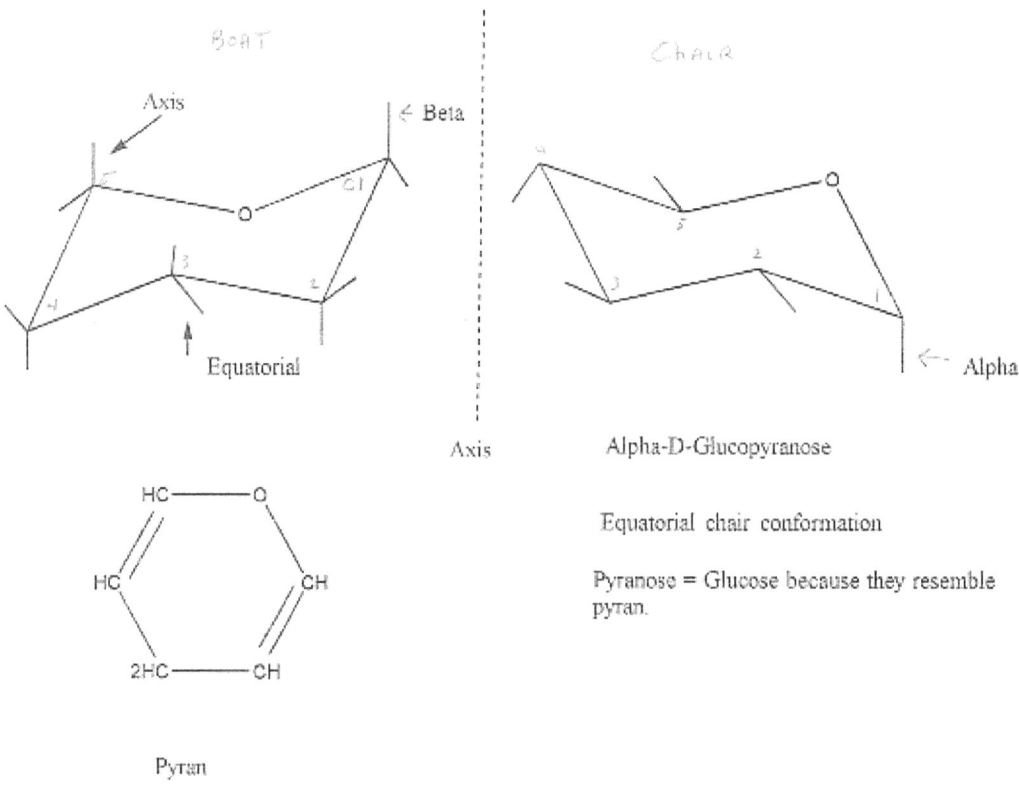

2 Chair conformations: Axis/equatorial orientation

BOAT

Axis

← Beta

Axis

Equatorial

CHAIR

Alpha-D-Glucopyranose

Equatorial chair conformation

Pyranose = Glucose because they resemble pyran.

Pyran

Compare to Haworth perspective formula

Alpha

Glucose was always considered a reducing agent. The H atom on C1 can dissociate and reduce another molecule. While the C atom becomes oxidized. A typical Oxidation-Reduction reaction of inorganic chemistry. With the use of Cu+2, you can test for the presence of a reducing sugar. If it is positive, there is a reducing end either as a aldehyde or hemiacetal. The enzyme that breaks down glycogen recognizes and starts at the non-reducing end of the polymer.

Reducing Agent/Non-reducing & reducing ends

The ring and open chain are in equilibrium

The open chain can be a reducing agent or become oxidized.

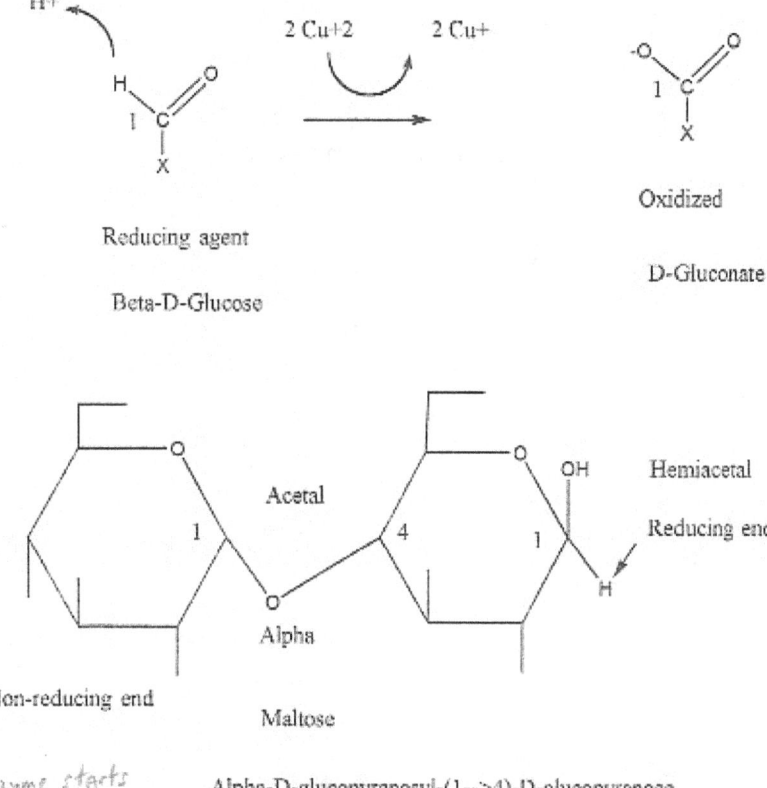

Reducing agent

Beta-D-Glucose

Oxidized

D-Gluconate

Acetal

Hemiacetal

Reducing end

Alpha

Non-reducing end

Maltose

Enzyme starts here!

Alpha-D-glucopyranosyl-(1-->4)-D-glucopyranose

The last comment for this detailed analysis of structure. The concept of chirality. A chiral center is a symmetric C atom. This means that all 4 covalent bond attachments are different. This leads to more isomeric forms. So the number of structures increase. To check for chirality, put the eraser end of a pencil on the C atom. Then look at the 4 attachments. Are they different? If yes, this atom is a chiral center and an isomer is possible. This check can be done in the open form or cyclic. Try it!

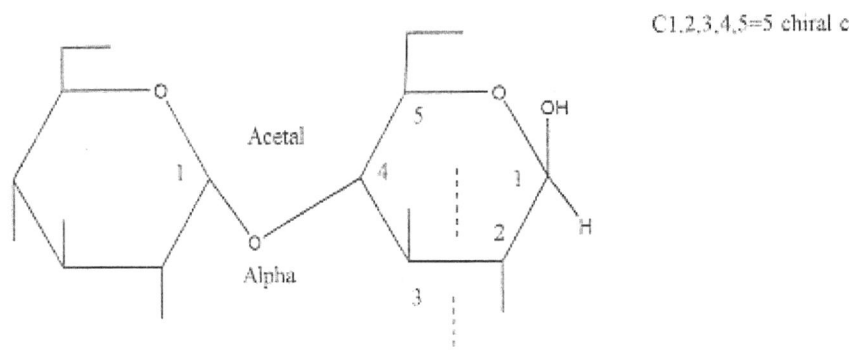

Chirality: asymmetric C

All attachments to C are different leading to isomeric forms

Straight chain form

C2,3,4,5=4 chiral centers

Single covalent bonds only qualify.

Check all 4 attachments are different.

Cyclic form

C1,2,3,4,5=5 chiral centers

Acetal

Alpha

The PPi moiety energizes the UDP-Glucose. This compound contains a large amount of potential energy. By the breakdown PPi, the total reaction has more than enough energy to transfer Glucose to the Glycogen polymer. PPi ensures that the reaction process proceeds forward. PPi is catalyzed by a hydrolysis process.

$$PPi + H2O \rightarrow 2 Pi \quad \textit{Pyrophosphatase}$$

Pi is inorganic phosphate.

Pyrophosphate PPi

on UDP-Glucose:
phospho diester link

H^+

H^+

H^+

H^+

on catalysis, H_2O is added by pyrophosphatase.

H^+ :OH

PPi energizes UDP-Glucose. PPi enables the transfer of Glucose to the Glycogen polymer.

3D
BALL-AND-STICK model

PPi Pyrophosphate

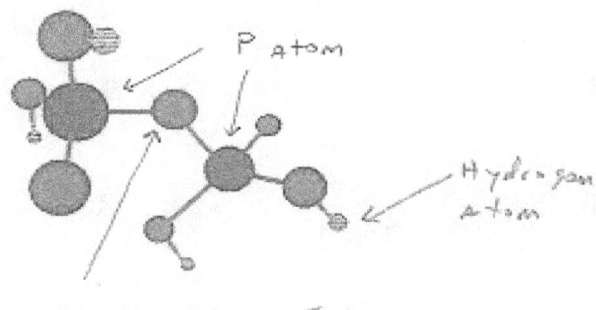

P atom

Hydrogen atom

oxygen atom Ester

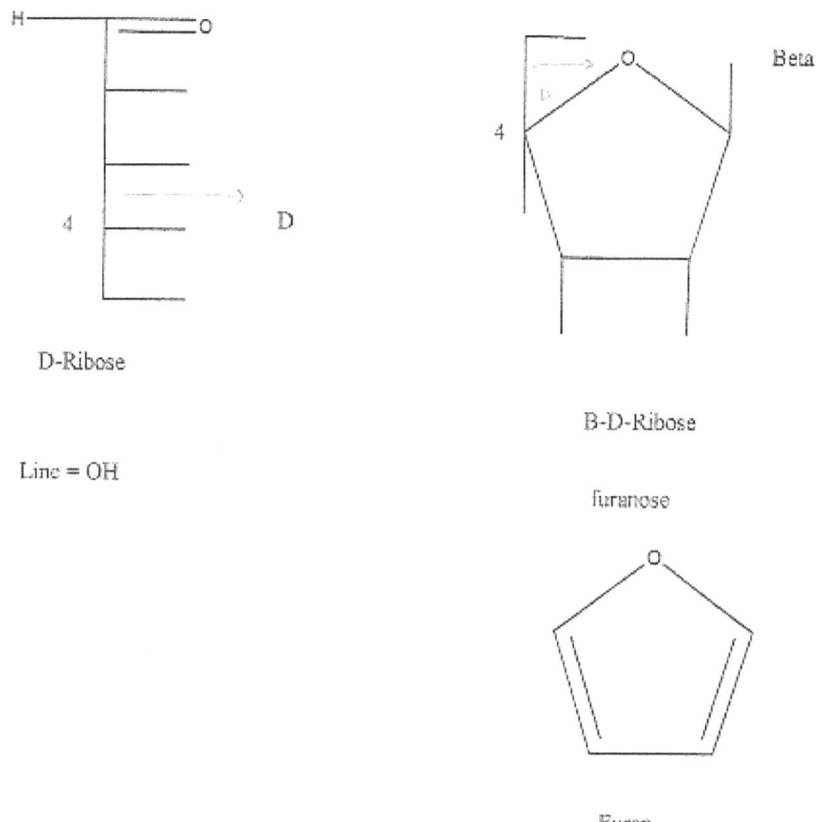

D-Ribose

Line = OH

B-D-Ribose

furanose

Furan

Note the likeness to Furan and naming the sugar a furanose. Same principles of structure applies. Ribose is a 5 C sugar, aldose.

3D
Ball-and-Stick Model

D-Beta-Ribose

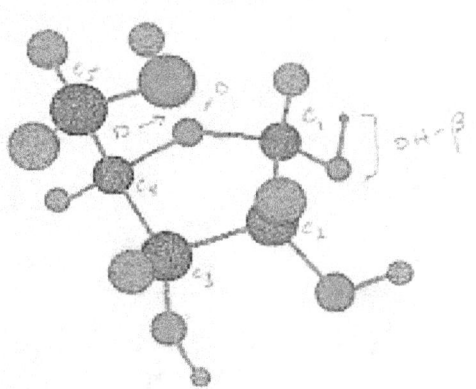

Uridine

Base

HN

Base

Base

Uracil
Nitrogen Base

Base

Base : accept a H+

2HC—OH

D

Beta

HC

CH

CH—CH

OH

OH

D-Ribose

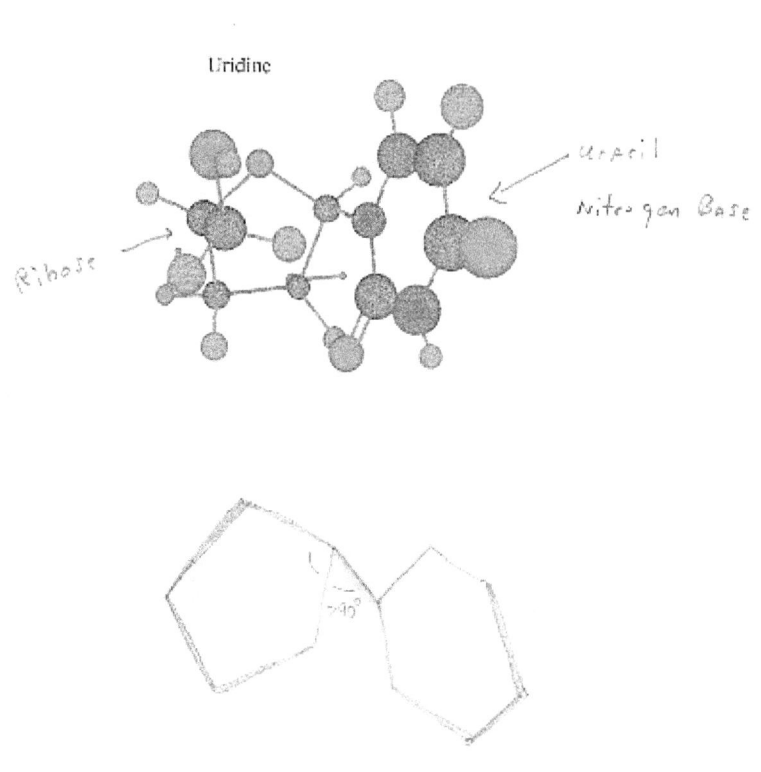

Uridine

Ribose

uracil

Nitrogen Base

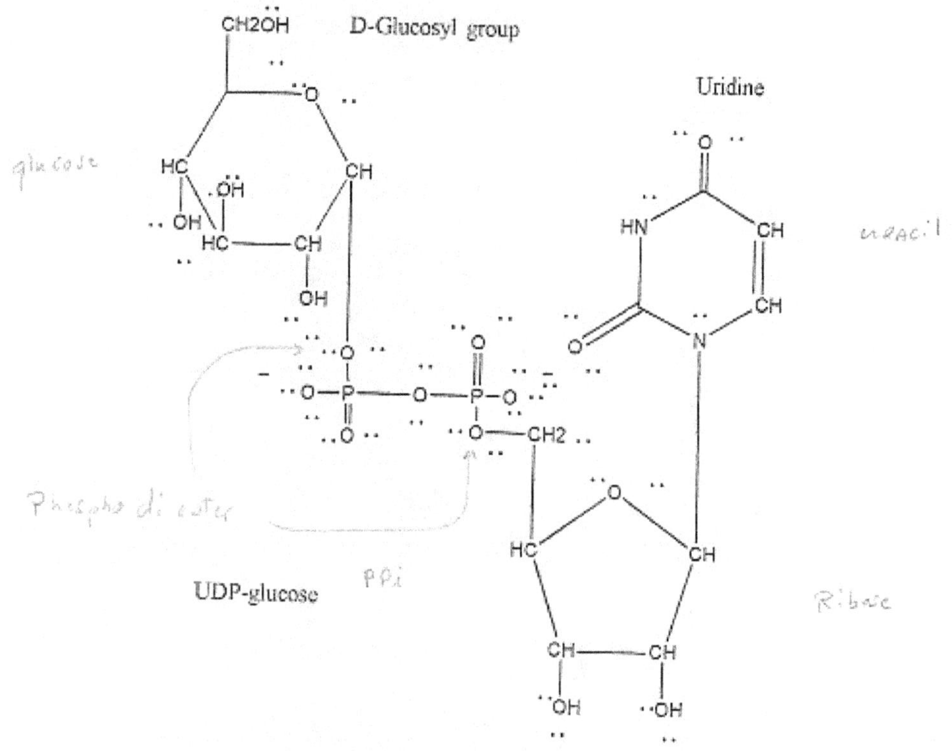D-Glucosyl group
Uridine
glucose
uracil
Phospho di ester
PPi
Ribose
UDP-glucose

UDP-Glucose

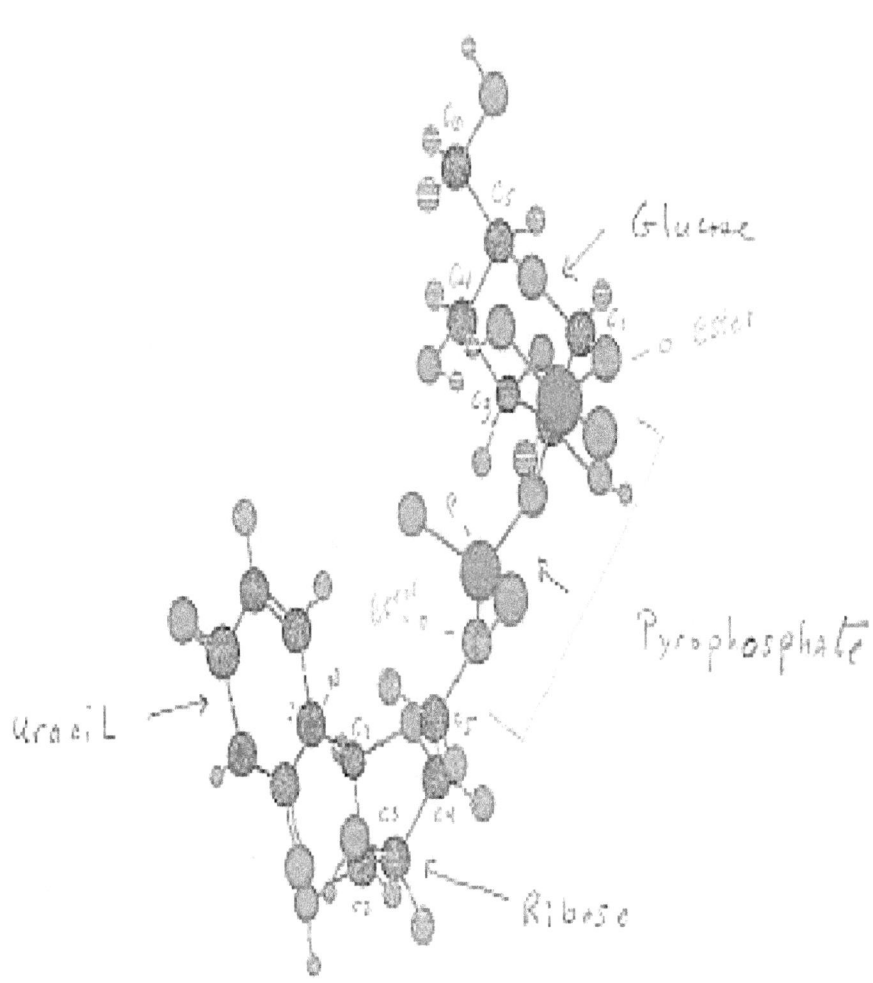

Glycogenin

We now turn to the second requirement before glycogen synthesis can start, Glycogenin. *Glycogen synthase* requires at least 8 sugar residues before it can start synthesizing glycogen. Glycogenin is a protein that acts as the primer and enzyme that starts this process.

      1. A nucleophilic attack by **Asp 162** creates a temporary intermediate with inverted confirguration.

      2. UDP-Glucose donates glucose to **Tyr 194** Hydroxyl group catalyzed by Glycogenin's intrinsic glucosyltransferase. This is a second nucleophilic attack restoring the starting configuration.

      3. Then chain-extending activity of the protein adds more residues in the same manner.

      4. Glycogenin retires and remains on the reducing end of the glycogen molecule. Glycogen synthase takes over.

There is a muscle Glycogenin and liver, *Glycogenin-2*.

$Mn+2$ is attached to the PPi moiety to stabilize the UDP leaving group. It is an electron-pair acceptor = Lewis Acid.

Tyr194 = Tyrosine (amino acid # 194 in chain).
Asp162=Aspartic acid (amino acid #162 in chain).
$Mn2+$ = cation = Lewis Acid=can accept electrons. 2 to be exact.
A leaving group is a functional group capable of dissociating easily from a molecule and it carries a negative charge with it.
Both attacks are nucleophilic types.
Lewis devised one of the concepts of acid base reaction using the idea of electron accepting and donating. So $Mn2+$ is an acid and UDP is a base.

      Nu: - $\rightarrow$ C 1 + (Partially + charge) = Polarized sigma bond. $\rightarrow$ Alpha-Glycosidic linkage.

      When all said and done, a mature glycogen particle has 12 tiers consisting of 55,000 glucose residues, 21 nm in diameter, and Mr of 10^7.

$nm = 10^{-9}$ m
Mr =relative molecular mass = [mass of a molecule of a substance/(1/12)C*12] dimensionless, only a ratio.

Glycogenin literally carries the carbohydrate primer for glycogen synthesis.
The Hydroxyl group on Tyr 194 is deprotonated be the nucleophilic attack occurs.

CH2OH

Nucleophilic

HC
OH
HC—CH
OH
OH

UDP-G

Sigma

O—P—O—P·O
O—Ribose—Uracil

CH2OH

HC
OH
HC—CH
OH
OH

Tyr 194

O—Glycogenin

Stays on reducing end

chain-extending activity

UDP

L.G. =Leaving group, prefers to dissociate.

6X

Glycogen Synthase

The location is the liver. From digestion, D-Glucose is converted to G-6-P.

D-Glucose + ATP → D-Glucose 6-phosphate + ADP

ATP, a high energy compound, is spent in the process. *Hexokinase IV*, glucokinase, is the enzyme.
  Some D-Glucose is converted by gluconeogenesis using the Cori cycle.
  G-6-P is converted to G-1-P by *phosphoglucomutase:*

G-6-P → G-1-P (Glucose 1-Phosphate)

G-1-P is converted to UDP-Glucose by *UDP-Glucose pyrophosphorylase*.

G-1-P + UTP → **UDP-glucose** + PPi

The branch points are made possible by the *amylo (1 → 4) to (1 → 6) transglycosylase*,
glycosyl (4 → 6) transferase.
  Branching increases **water solubility** and the **number of non-reducing ends**.

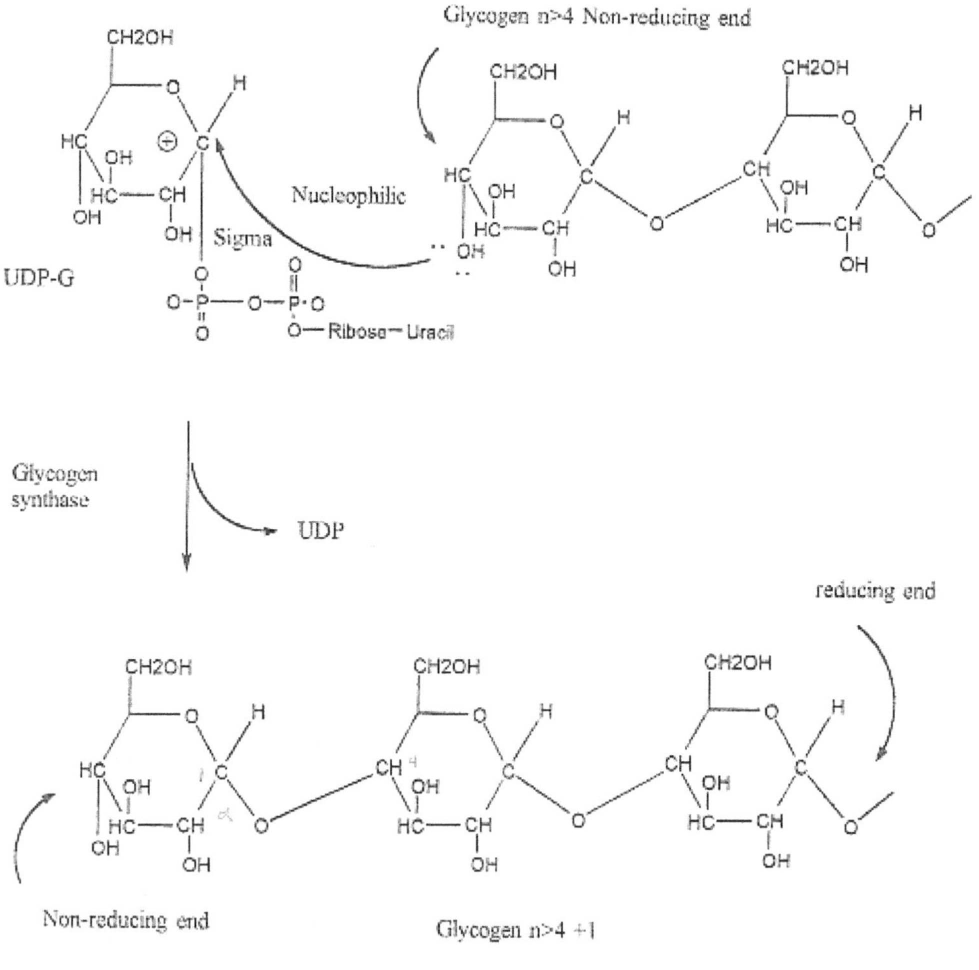

Glycogen n>4 Non-reducing end

UDP-G

Nucleophilic

Sigma

O—Ribose—Uracil

Glycogen
synthase

UDP

reducing end

Non-reducing end

Glycogen n>4 +1

40

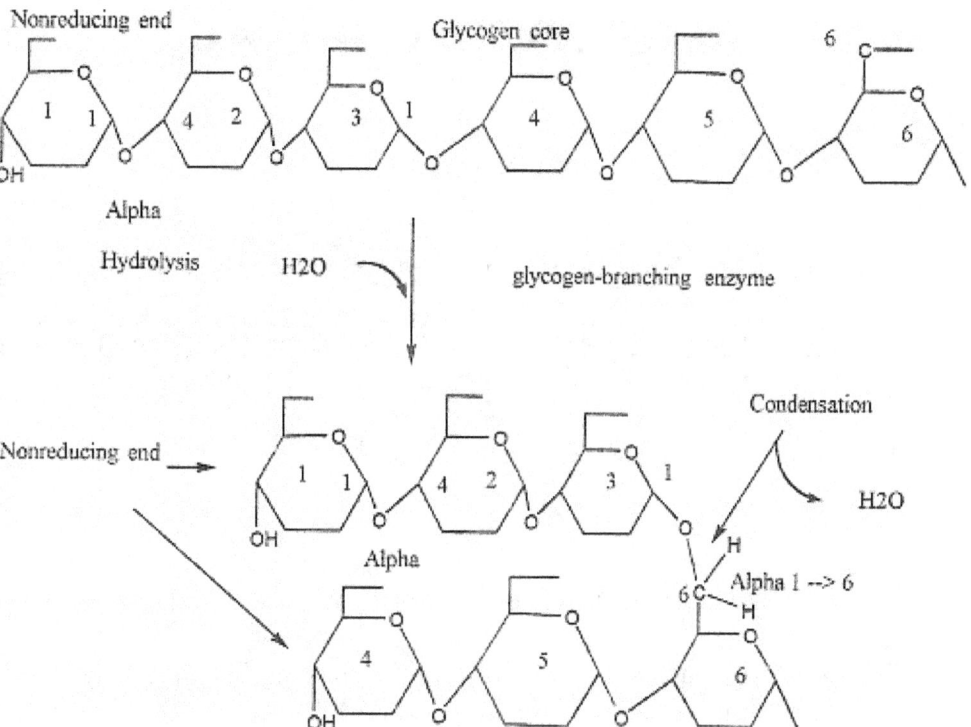

Nonreducing ends increased

Alternating Hydrolysis and Condensation rxs.

Chapter 4

Glycogen Breakdown

Overview

Glycogen is a very large polymer. 3 enzymes start on the outer branches of glycogen.

1. *Glycogen phosphorylase* : Starts at the non-reducing end. (alpha 1 → 4) glycosidic linkage is attacked by Pi producing Alpha-D-Glucose 1-phosphate.

Glycogen + Pi → Alpha-D-Glucose 1-phosphate

Amylase catalysis of glycosidic linkage during digestion is a hydrolysis reaction. Water is added to the molecule.

G-1-P retains some of the energy in the phosphate moiety. In other words, G-1-P is reactive and energized.

B6, pyridoxal phosphate, is the cofactor for this enzyme. It's phosphate group acts as a general acid catalyst, promoting attack by Pi on the glycosidic bond.

The enzyme repeats the process to 4 residues from (alpha 1 → 6) branch point.

2. Debranching enzyme, *oligo (alpha 1 → 6) to (alpha 1 → 4) glucantransferase* : There are 2 successive transfers of branches and a hydrolysis of C6.

3. *Phosphoglucomutase* : A double phosphate group transfer by the Ser residue of the enzyme.

G-1-P → G-6-P

In muscle, G-6-P is used for energy during muscle contraction.

In the liver and kidney, G-6-P is converted to D-glucose by *glucose 6-phosphatase*. This provides blood glucose after fasting for organ function, especially brain.

Muscle and adipose tissue do not possess this enzyme.

Muscle is depleted of their glycogen stores in 1 hour during exercise.

Liver is depleted in their stores in 12-24 hours.

Vitamin B6 Acid Catalyst

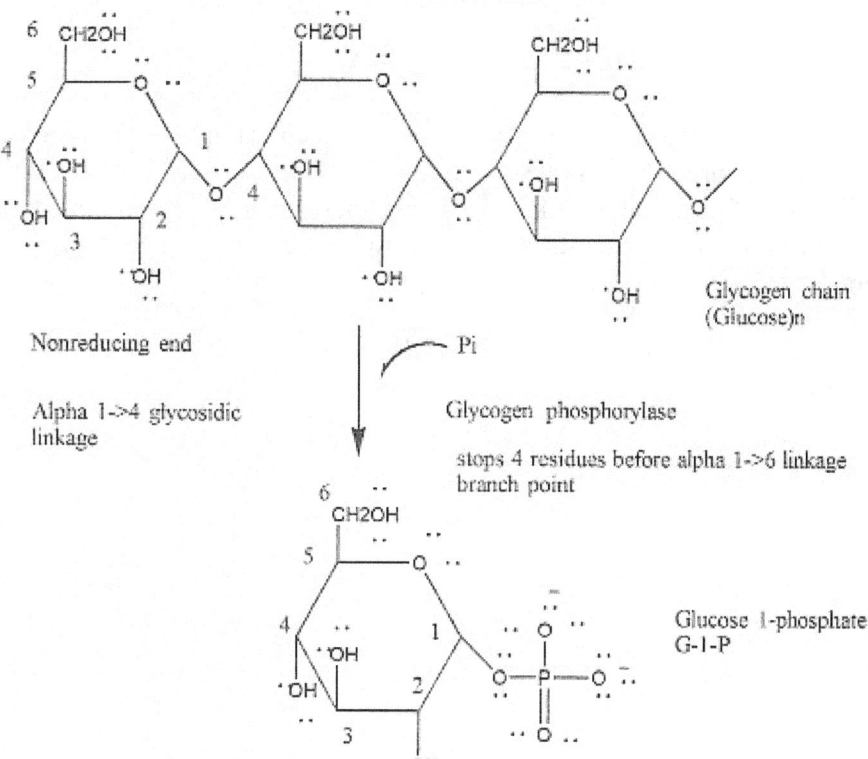

Glycogenolysis: Breakdown from glycogen to G-1-P

Nonreducing end

Glycogen chain
(Glucose)n

Alpha 1->4 glycosidic
linkage

Pi

Glycogen phosphorylase

stops 4 residues before alpha 1->6 linkage
branch point

Glucose 1-phosphate
G-1-P

G-1-P -->1.G-6-P-->Glycolysis-->ATP-->work
or -->2.liver-->blood glucose--CNS or muscle
glycogen.

+

(Glucose)n-1

Nonreducing end

Vitamin B6 is the cofactor to the enzyme glycogen phosphorylase. There are several classifications for cofactors. A prosthetic group type is covalently bonded to the enzyme. The coenzyme type is loosely-bound. The cofactor helps the enzyme in it's biological activity. Pyridoxime is phosphorylated by ATP to Pyridoxal phosphate. When an inactive enzyme, apoenzyme, is completed by a cofactor, it is called a holoenzyme.

Vitamin B6 acts as a Lewis acid catalyst. It is not changed by the reaction process. A Lewis acid donates a H+.

$$\text{Pyridoxime} + \text{ATP} \rightarrow \text{PLP} + \text{ADP}$$

The location for this process is the liver. Where Glycogen is degraded to glucose to restore blood glucose levels for vital organ functions, eg. Brain. The 2 main signals to start this process is Glucagon and Epinephrine. 2 hormones which call for an increase in glucose. Glucagon is opposite of Insulin and Epinephrine is opposite of nor-epinephrine. Intense exercise raises epinephrine and a falling glucose from the same stress raises Glucagon.

The debranching enzyme has 2 activities:
1. Transferase activity.
2. Glucosidase activity which performs a hydrolysis.

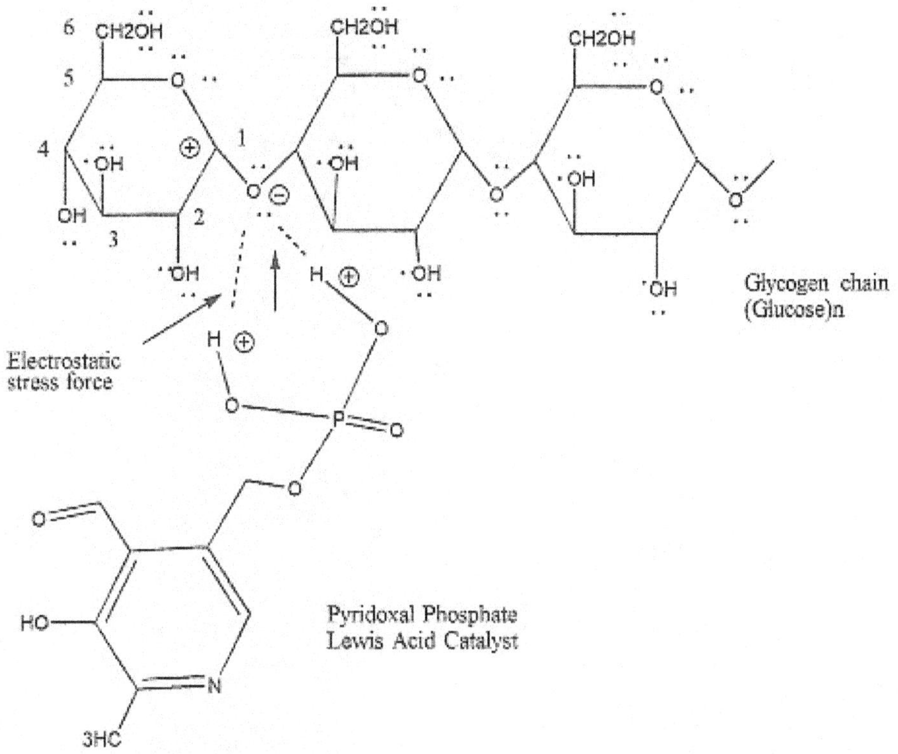
6
CH2OH
5
4
OH
3
2
OH
1
OH
CH2OH
OH
OH
CH2OH
OH
OH
Glycogen chain
(Glucose)n
Electrostatic
stress force
H
H
P
O
HO
N
3HC
Pyridoxal Phosphate
Lewis Acid Catalyst

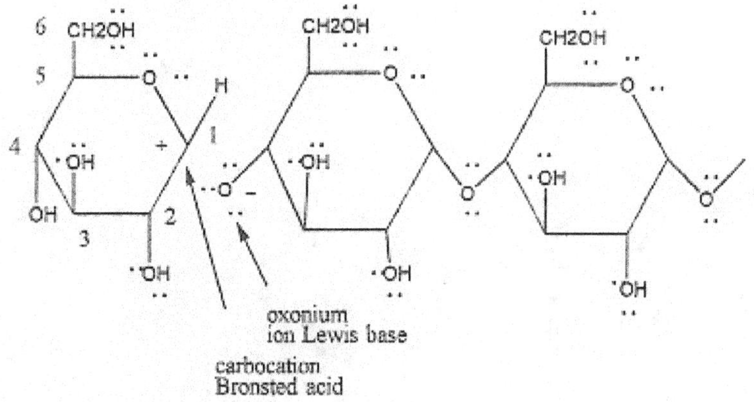
6
CH2OH
5
H
4
OH
1
OH
2
3
OH
CH2OH
OH
OH
CH2OH
OH
OH
oxonium
ion Lewis base
carbocation
Bronsted acid

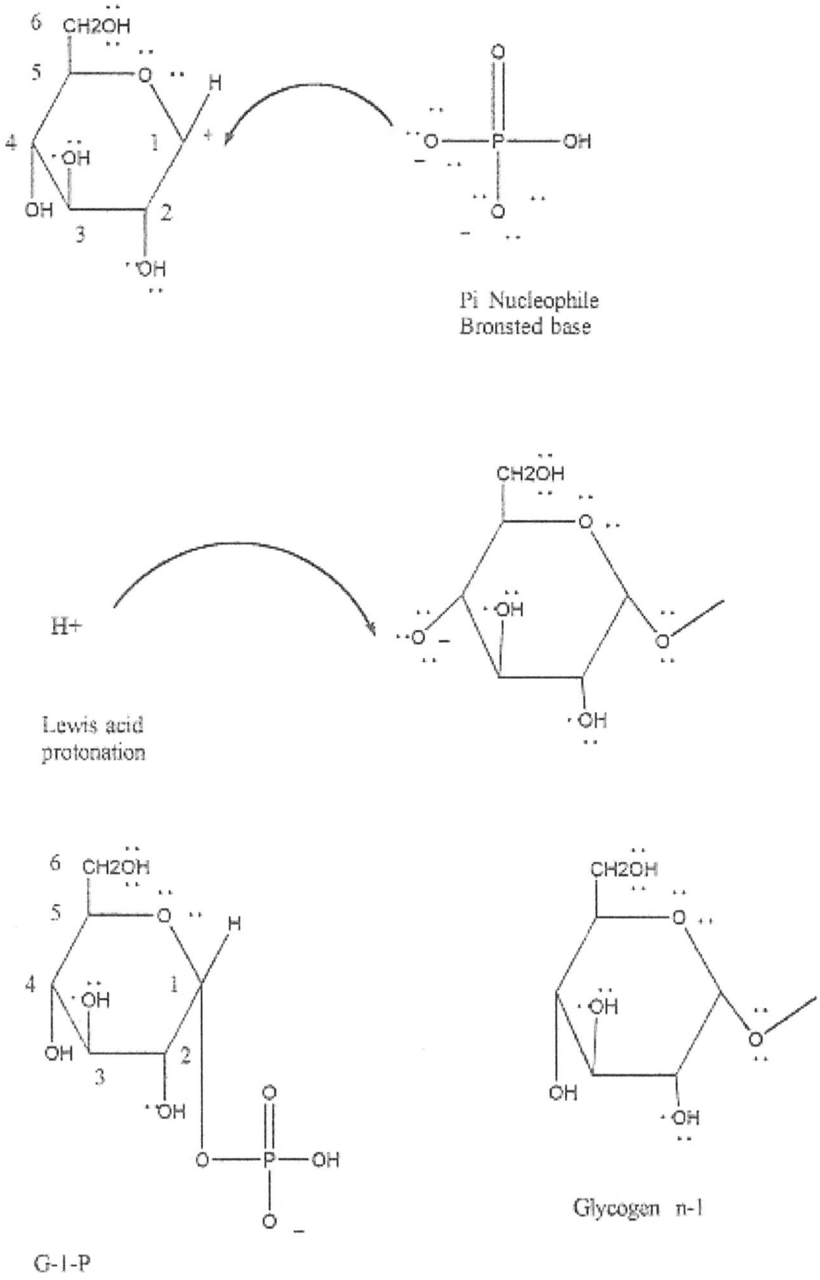

Pi Nucleophile
Bronsted base

Lewis acid
protonation

H+

G-1-P

Glycogen n-1

Nonreducing end

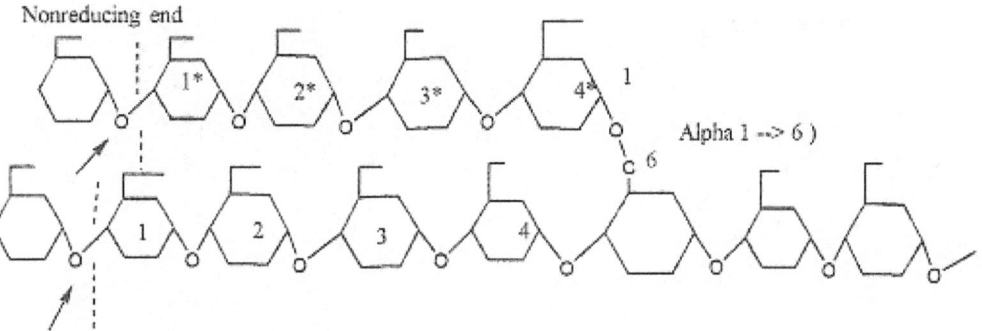

Glycogen phosphorylase stops 4 residues before (alpha 1 --> 6)

transferase activity of debranching enzyme

(alpha 1 --> 6) glucosidase activity of debranching enzyme

Glycogen phosphorylase catalyzes unbranched polymer
Debranching enzyme catalyzes the branches and feeds phosphorylase.

Debranching enzyme (alpha 1-->6) glucosidase

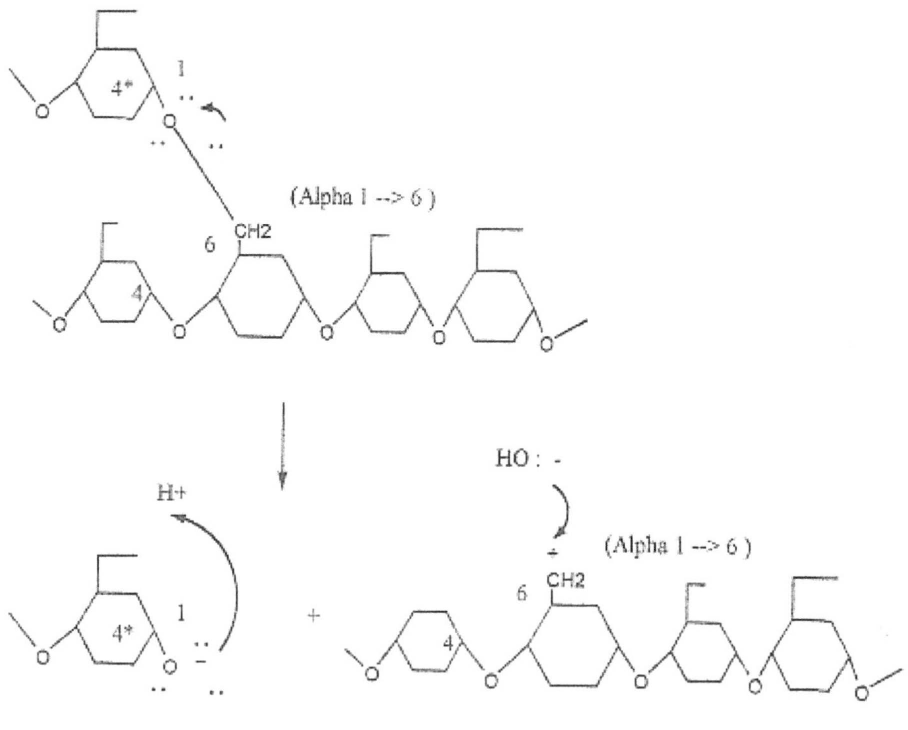

(Alpha 1 --> 6)

CH2

Hydrolysis : Addition of water

Phosphoglucomutase

G-1-P can enter glycolysis to produce energy for work, muscle contraction. Or it can restore blood glucose, a liver function. G-1-P is converted to G-6-P by *phosphoglucomutase.*

$$G\text{-}1\text{-}P \rightarrow G\text{-}6\text{-}P$$

The SER residue of the enzyme performs a double phosphate transfer on the same glucose molecule.

1. PO4-2 to C6.
2. Remove PO4-2 from C1.

The enzyme _Glucose 6-phosphatase_ is found in the kidney and liver. It is an integral membrane protein of the Endoplasmic Recticulum with it's location on the lumenal side. The G-6-P is transported into the E.R. and degraded to Glucose and Pi. Both transported back to the cytosol by 2 different transporters. A second tranporter moves the glucose from the hepatocyte to the plasma.

$$G\text{-}6\text{-}P \rightarrow Glucose + Pi$$

By compartmentalizing the Phosphatase enzyme, the process can control the fate and metabolism of glucose.

Muscle and adipose tissue do not possess this enzyme.

Pi is a good leaving group. This means it likes to dissociate from the parent molecule. The Hydroxyl group is a nucleophile which attacks the C6 atom of Glucose. The Pi is protonated. Both the H+ and OH – ions come from water dissociation. Water is added. Can you write the reaction mechanism of Glucose 6-phosphatase?

With Glucose created, it is transported by 2 membrane proteins, T2 and GluT2.

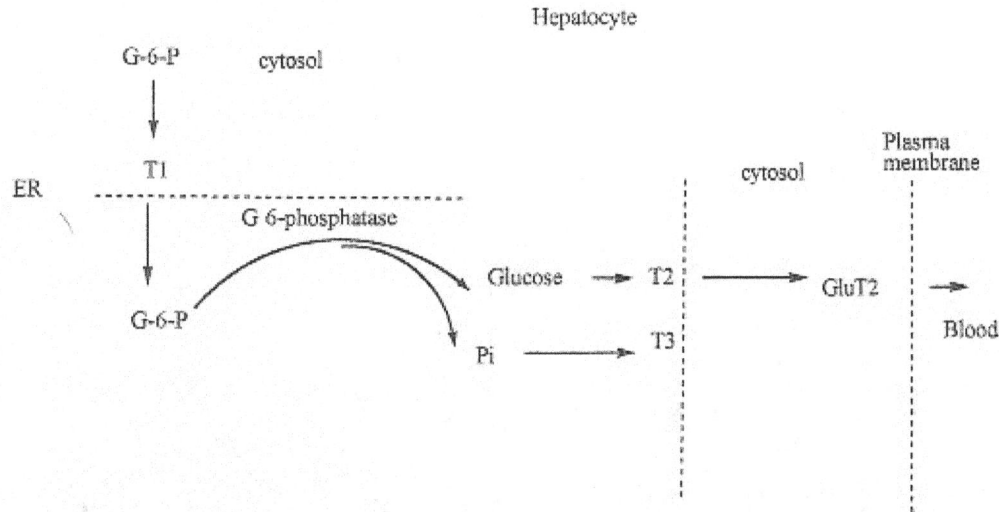

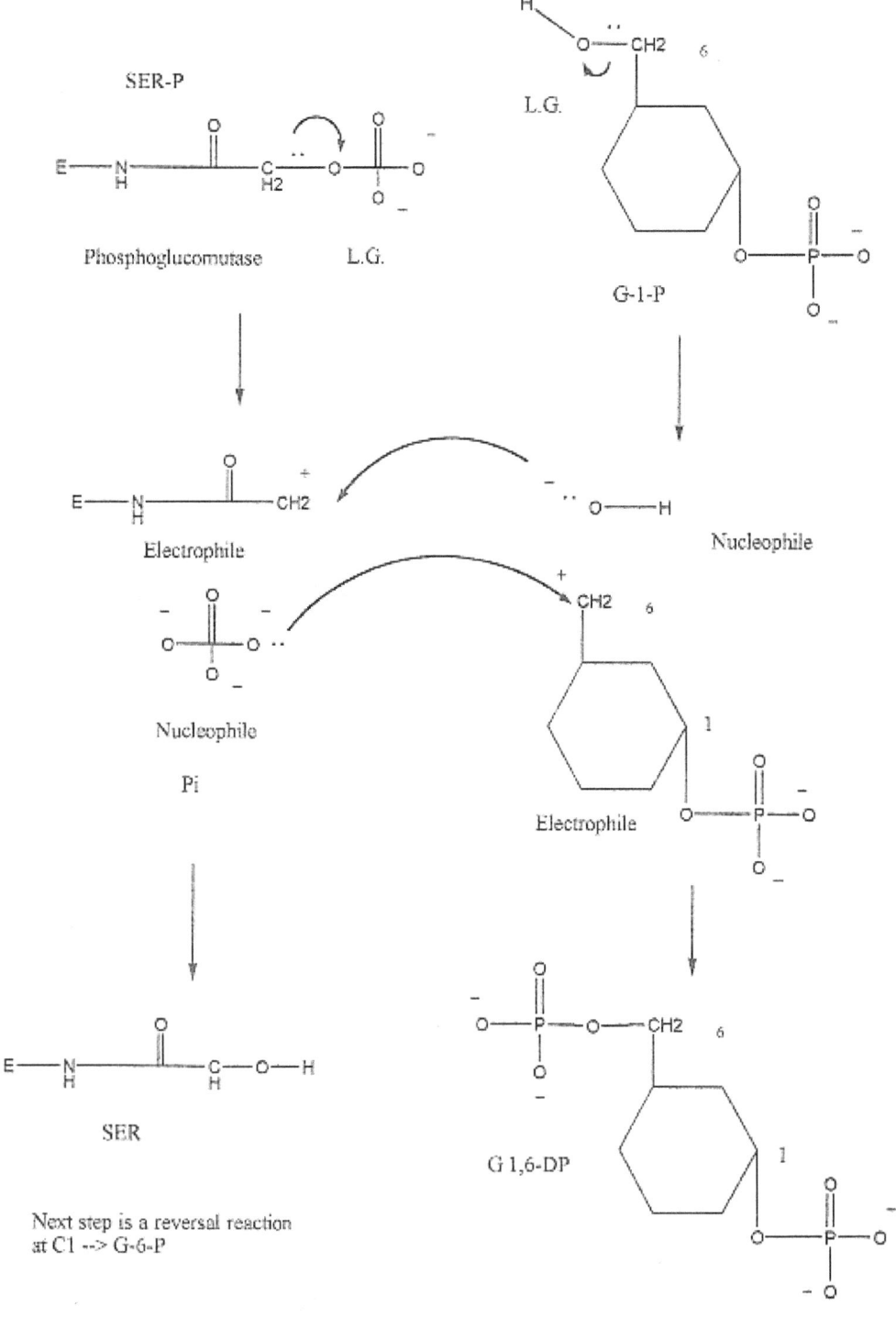

Can you write the reversal reaction? What is regenerated? Remember a catalyst does not change in the reaction.

The enzyme is energized by ATP originally. In the mutase reaction, there is no change in matter or energy. There is only change in configuration, structure.

Chapter 5

Lipids

Digestion

Most of the fat in the diet consists of triglycerides (neutral fats). They are neutral because they are no charged, + or -. Triglycerides are made of glycerol and 3 fatty acids. Their synthesis is a process of condensation in which 3 water molecules are removed in the the process. Like carbohydrates, digestion is the process of hydrolysis. Water is added to the triglycerides with the return of the starting materials, triglycerides and 3 fatty acids. Neutral fats are found mainly in food of animal origin.

H2C-<u>OH</u>
  |
HC-<u>OH</u>   Glycerol, a 3C alcohol
  |
H2C-<u>OH</u>

3   <u>HO</u>-(C=O)-(CH2)16-CH3   Fatty acid, a carboxylic acid

     HO-(C=O)-R   R=Alkyl group, a hydrocarbon chain

The underlined atoms are the components of water that are added during hydrolysis and removed during condensation.
Other fats present in the diet are:
Phospholipids are triglycerides with 1 fatty acid attached to glycerol by a phosphate group ether linkage.

R-O-R'   ether

Phospholipid

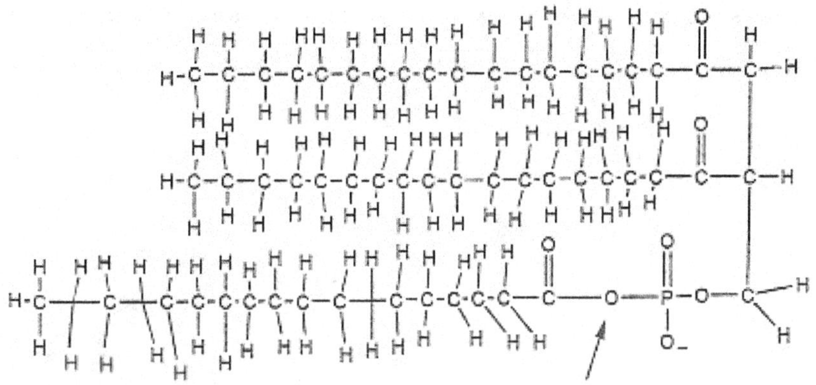

ether linkage

Note the ether linkage:

R-(C=O)-O-(P=O)-R'

Also the – charge on the oxygen atom of the phosphate group.

Cholesterol is a steroid nucleus type lipid. It's structure is not like that of a triglyceride. But it is handled as a lipid in biological systems.

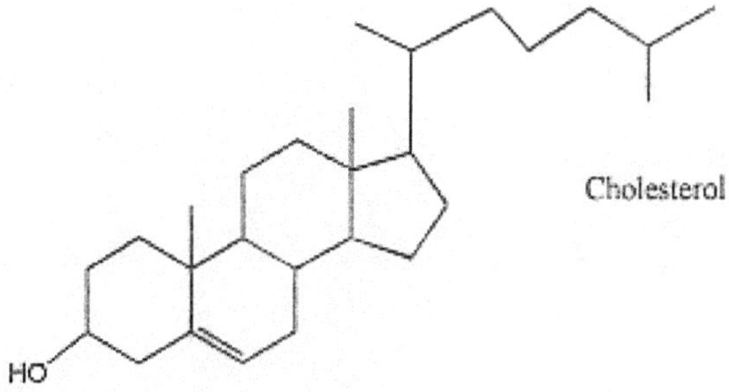

Cholesterol

nucleus

Note the alcohol group on the lower left ring.

There is only 1 double bond in the structure, which is the only unsaturated area in the molecule. Saturation means that the carbon atom has 2 or 3 H atoms attached to it.

At the end of a line or bend, there is a carbon atom with it's appropriate amount of H atoms.

Cholesterol esters are cholesterol with an esterified fatty acid group attached to it.

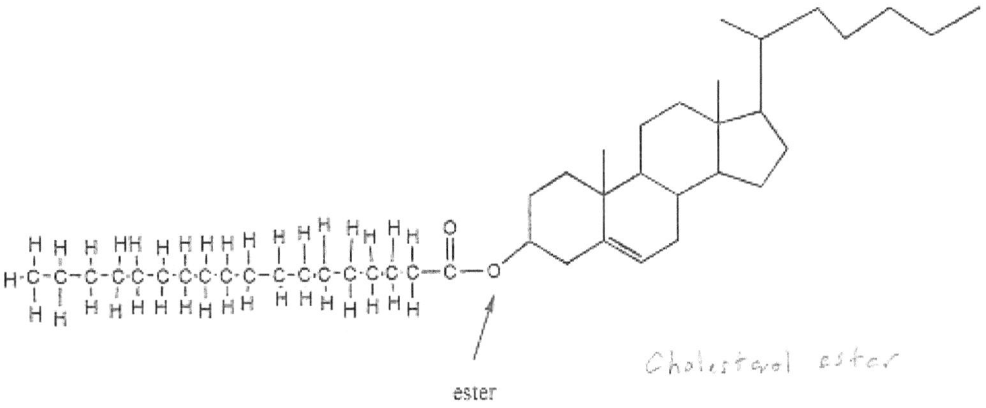

ester

R-(C=O)-O-R'

Cholesterol ester

When you look at triglyceride, you see a large amount of potential energy in the long saturated hydrocarbon chains. It looks like a ketone:

R-(C=O)-R'

In the digestion process, the products look like an aldehyde:

R-(C=O)-H

Or they look like an alcohol:

R-OH

But the products are carboxylic acids. This is due to the fact that the aldehyde is oxidized further to the carboxylic acid product which is the starting material, fatty acid.

At the beginning, a triglyceride is synthesized by the alcohol, glycerol and carboxylic acid, fatty acids. There is a condensation reaction with the elimination of 3 molecules of water in the process. The reverse is digestion. The biological system, through digestion, simply reverses condensation by hydrolysis and end with the starting materials, alcohol and carboxylic acid.

In the following figures, it shows that a H: - , Hydride, is produced in digestion. This specie does not really exist. It is very highly reactive. Also H2, a gas, is produced. Biological systems won't tolerate this product. So the oxidation process is done with a different mechanism to obtain the carboxylic acids.

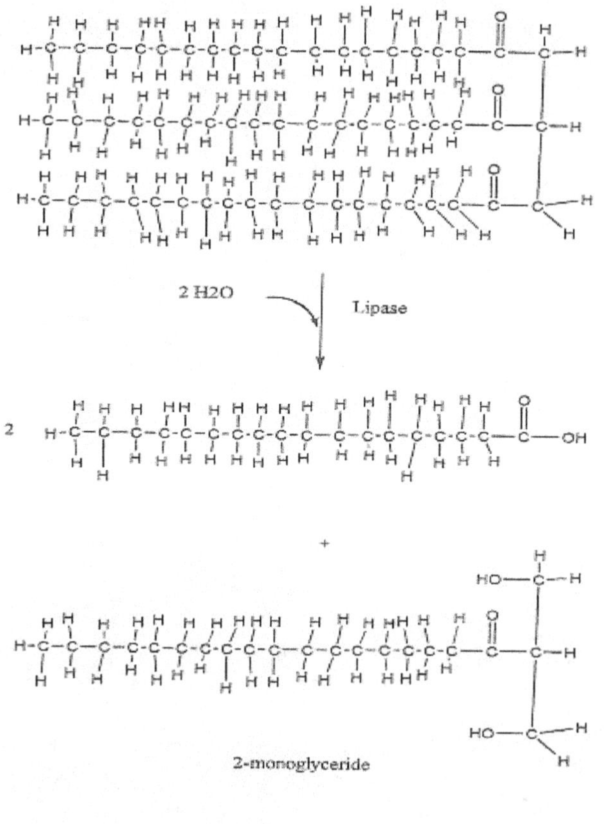

Tristearin : C18

2 H2O — Lipase

Tristearin : C18

2-monoglyceride

Lipase found in the saliva and pancreatic fluid splits triglyceride into a monoglyceride and 2 fatty acids with the addition of 2 water molecules.

Most of the fat is digested in the small intestine by the emulsification process with bile acids and Lecithin.

Stomach agitates the fat globules and breaks them into smaller ones. The digestive enzymes are all water soluble; so they operate only on the outside of the globule.

The bile salts, a steroid nucleus type fat and lecithin, a phospholipid, are both fat and water soluble. Both also carry a negative charge and lecithin has also a + charge. By mixing with the fat globule, they lower the interfacial tension of the fat. This exposes the fat molecules to the water soluble digestive enzymes by fragmentation. This also increases the surface area of the globules to further enhance the enzyme activity.

The bile salts have 2 additional functions: 1. They form micelles which increase the water solubility of the fat globule and 2. They, "ferry", transport the digestate to the brush border for further digestion.

This is a good example of like dissolving like. Oil dissolves oil. But what about being in a water soluble environment. The answer is amphoteric molecules, part oil and part water soluble.

To be water soluble, a molecule needs to carry a charge. The enzymes are proteins with side functions which carry a lot of charges.

Micelle formation is spontaneous due to the natural separation of water and oil.

Bile salts are considered salts due to the ionized nature of bile acid with the association with Na+. It is a simple acid base reaction with the Na+ coming from the base.

Here are some points to remember about the lipids involved in the emulsification process.

1. Fatty acid and 2-monoglyceride are products of the digestive process.
2. Triglyceride, cholesterol, cholesterol ester, and phospholipids are reactants in digestion.
3. Bile acid is a carboxylic acid with a cholesterol, steroid, nucleus . It is more water soluble because of the Alcohol groups it possesses.
4. Lecithin is a phospholipid with a charged phosphate group attached.
5. Covalent single bonds can rotate and bend slightly.

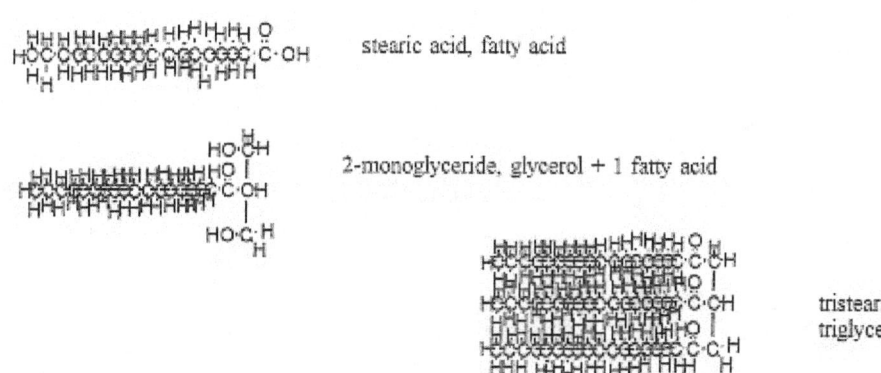

stearic acid, fatty acid

2-monoglyceride, glycerol + 1 fatty acid

tristearin, triglyceride

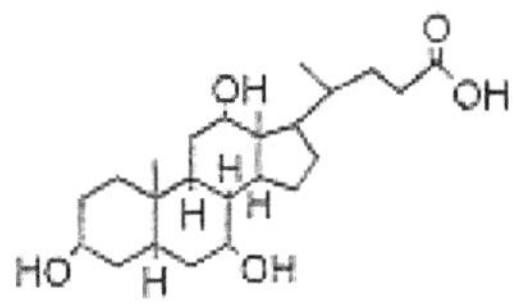

cholic acid, bile acid

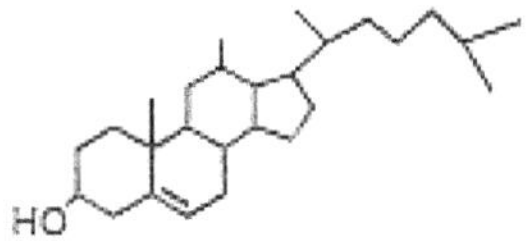

Cholesterol, steroid lipid

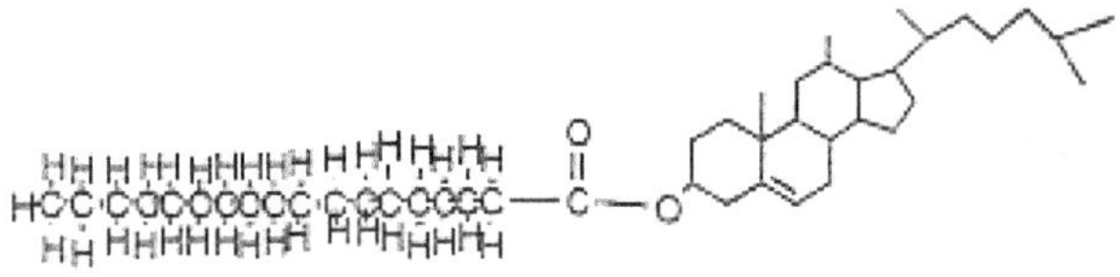

Cholesterol ester, cholesterol + 1 fatty acid

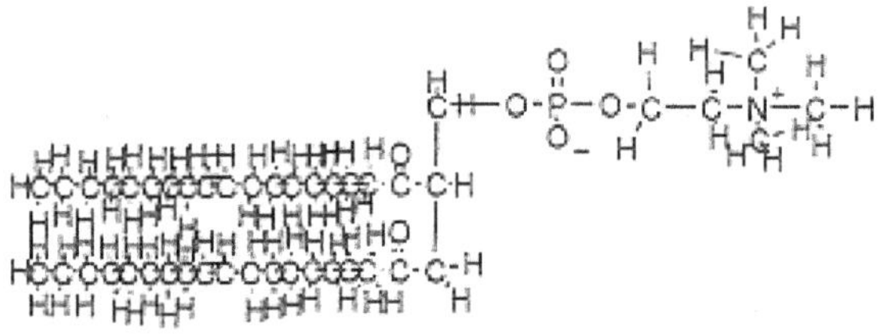

Lecithin, phospholipid

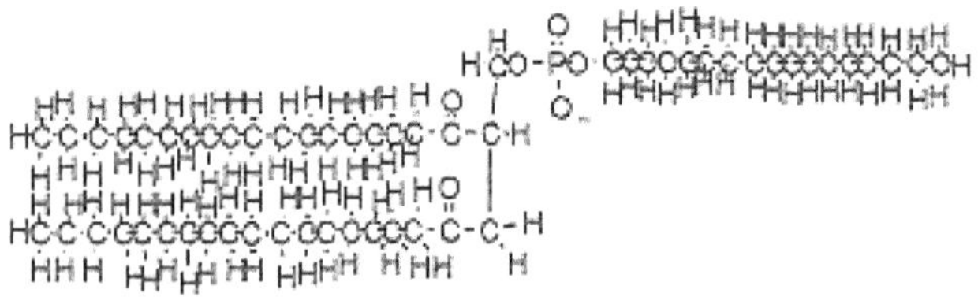

Phospholipid

Emulsification

A large fat globule consisting of un-digested fat

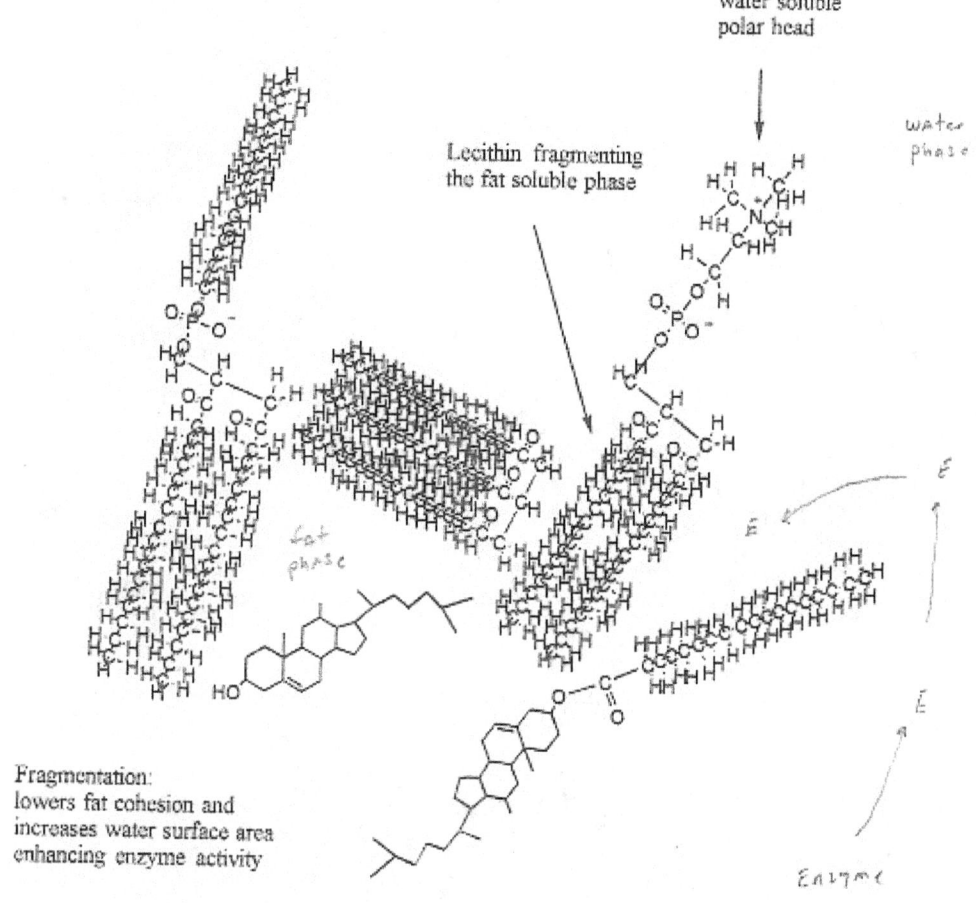

water soluble
polar head

Lecithin fragmenting
the fat soluble phase

water phase

fat phase

E

E

E

Enzyme

Fragmentation:
lowers fat cohesion and
increases water surface area
enhancing enzyme activity

Emulsification

A small fat globule consisting of digested and undigested fat

digestates being generated and leaving globule

E

Enzyme

water soluble polar head

D   D

Lecithin fragmenting the fat soluble phase

D

D

D

D

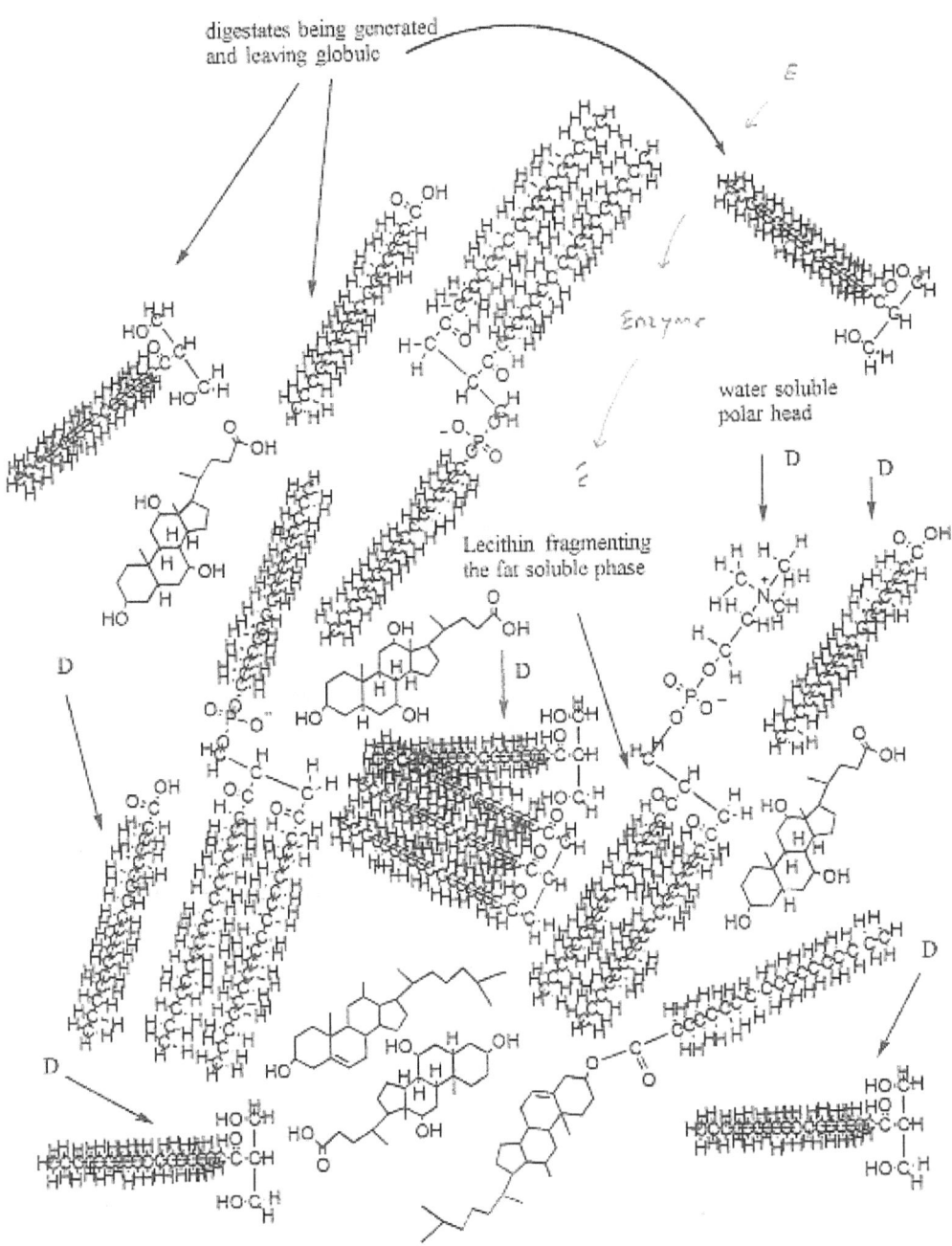

Absorption

Bile acid micelle consisting of digestates
The "ferry"

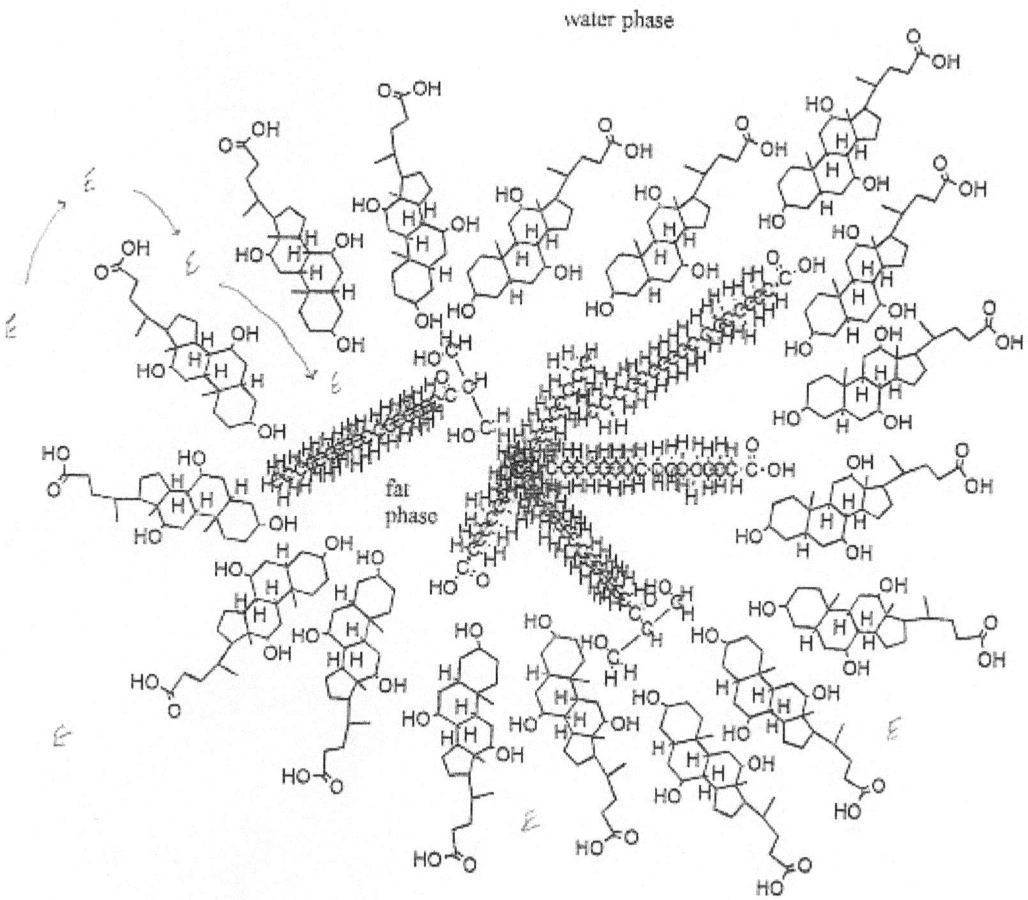

water phase

fat
phase

Enzyme

Brush Border
microvilli
Small Intestine

As already mention above, the ferry brings the digestates to the epithelial cells of the brush border. Here they are taken up by pinocytosis and brought to the Smooth Endoplasmic Recticulum. The digestates are once again made into new triglycerides and released at the base in the form of chylomicrons. These chylomicrons are lipid and protein micelles that transport the lipids in the blood to their final destination. This path is via the thoracic duct to the general blood circulation.

Since small and medium chain lipids are more water soluble, some are taken up directly into the portal blood supply.

Chapter 6

Proteins

Digestion

Like the other 2 main foods, proteins are digested by hydrolysis. Proteolytic enzymes split the amino acids in proteins by the addition of water. The reverse process is called condensation which joins to amino acids by the withdrawal of water. The area involved is called the peptide linkage.

Amino terminus 　　　　NH3+　　　H
　　　　　　　　　　　　 |　　　　　 |
　　　　　　　　R-CH-(C=O)-N-CH-COO- Carboxyl terminus + H-OH
　　　　　　　　　　　　　　　　　 |
　　　　　　　　　　　　　　　　　 R

There are 2 ends: Amino = NH3+ and Carboxyl = COO-
Amino has base character and Carboxyl has acid character.
The peptide linkage is in bold : a carboxylic acid derivative.
The structure is obviously a condensation product with water removed.
R is the side functions of the amino acids.

The reaction mechanism is a nucleophilic and electrophilic attack on the peptide linkage.

Protein Hydrolysis

Proteolytic enzyme

Carbocation

amino acids are polar = charged
because of pH and water environment.

Digestion of the protein starts in the stomach. Pepsin is the lead enzyme. This enzyme requires a low pH for optimal function. The parietal cells of the gastric gland secrete H-Cl to achieve this pH. The main function of Pepsin is the digestion of collagen, an intercellular connective tissue of meats. It is like albumin in structure. The net result is protein converted to proteoses, peptones, and a few polypeptides.

The next stage of digestion is in the upper small intestine. The pancreas secretes Trypsin, Chymotrypsin, Carboxy-polypeptidase, and Proelastase.

Trypsin and Chymotrypsin → small polypeptides.

Carboxy-polypeptidase → attack COO- ends of polypeptides.

Proelastase → Elastase digests elastin fibers.

Net digestion is a few amino acids and mostly dipeptides and tripeptides.

In the small intestine at the brush border, microvilli, are peptidases associated with the enterocytes. 2 important examples of enzymes are aminopolypeptidase and dipeptidases. Inside the enterocyte are other peptidases that finish the job. Most of the final digested products are amino acids.

Protein → amino acids

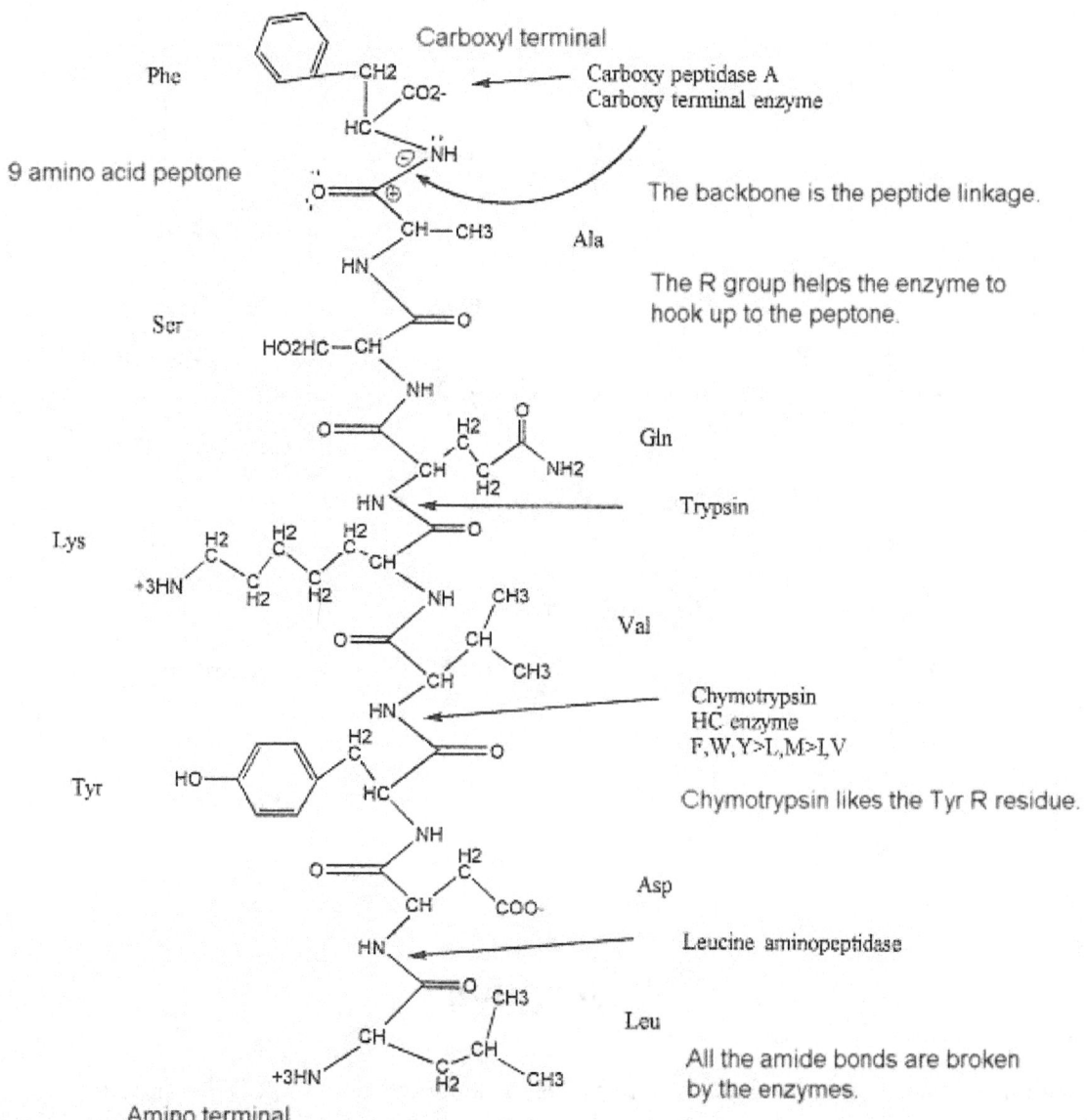

Carboxyl terminal

Phe

Carboxy peptidase A
Carboxy terminal enzyme

9 amino acid peptone

The backbone is the peptide linkage.

Ala

The R group helps the enzyme to
hook up to the peptone.

Ser

Gln

Trypsin

Lys

Val

Chymotrypsin
HC enzyme
F,W,Y>L,M>I,V

Tyr

Chymotrypsin likes the Tyr R residue.

Asp

Leucine aminopeptidase

Leu

All the amide bonds are broken
by the enzymes.

Amino terminal

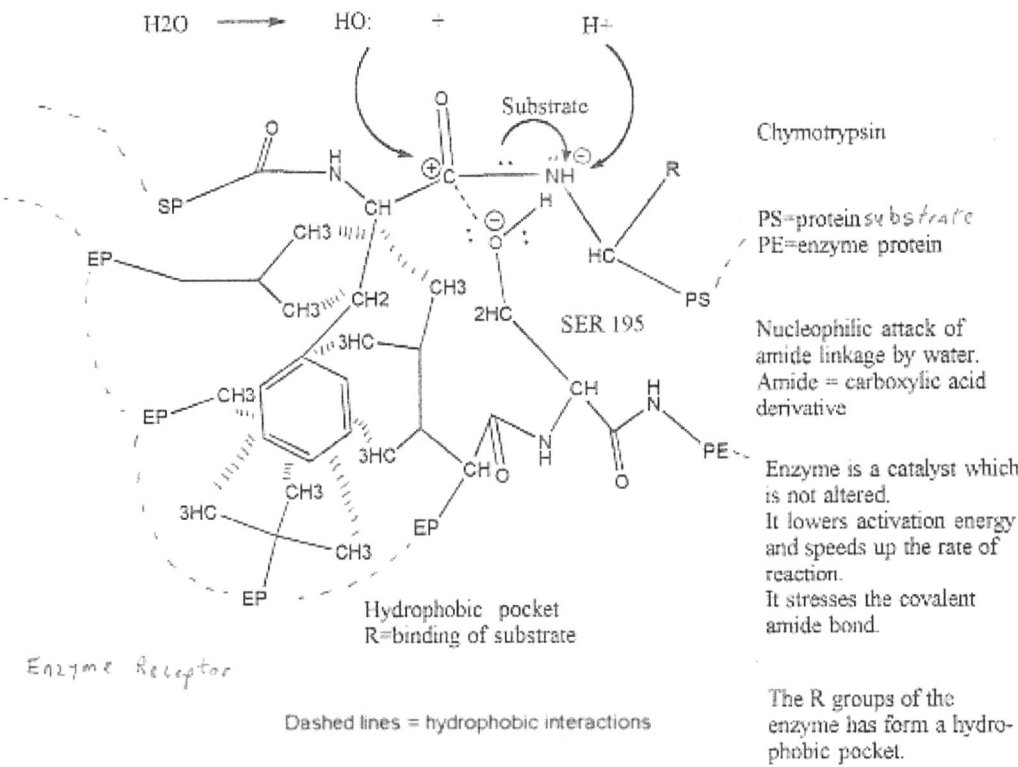

The enzyme causes enough stress to allow water to attack the protein molecule.

Absorption

With the final digested products of dipeptides, tripeptides, and amino acids present, the cell transports them by co-transport (secondary active transport). This is facilitated by a transport protein and Na+. This is similar to the transport of glucose. A few amino acids are transported like fructose, by facilitated diffusion. There is only a transport protein with no Na+ assist. As you might think, there are a multitude of transport proteins for the given amino acids.

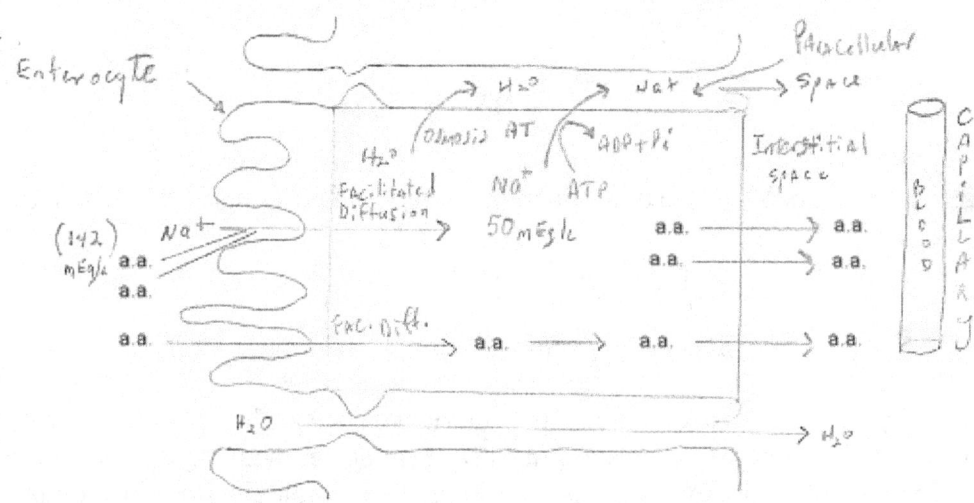

You have seen this figure before; but instead of the sugars, amino acids are shown.

Chapter 7

Energy

We have arrived at the point where our 3 main nutrients have been digested and absorbed.

Carbohydrate → Glucose
Fat → Triglyceride
Protein → Amino Acid

The football player consumes a large amount of nutrient calories while playing the game. This exercise is mainly aerobic in nature. This means that catabolism occurs with oxygen as the final electron acceptor. Then with protonation, H_2O is the end product. Biological combustion products are CO_2 and H_2O. Under aerobic conditions, the nutrient generates the maximum amount of energy required. It is also the most efficient of the 2 conditions. The player can experience an anaerobic state producing fatigue, mental dullness, and muscle cramping. This is likely to occur because of poor preparation for the game such as hydration, meals, warming up and etc. Anaerobic metabolism leads to lactic acid build up.

Metabolism → Anabolism = creation = potential energy
→ Catabolism = destruction = kinetic energy → work

Biological systems have the following preferences for energy production.

Glucose > Triglyceride > Amino acid

Glucose is the most universal and preferred substrate for energy production. The triglyceride is usually stored as potential energy for periods of starvation. Also glucose and amino acids can also be stored the same way if activity is very low. Amino acids are mainly used for anabolic purposes such as tissue replacement, immune activity, hormone and nervous function.

In our case, the football player, glucose and triglyceride are used tor energy. The amino acids for anabolism. The triglyceride is probably used first, before glucose, because glucose is in the the form of glycogen which requires glycogenolysis to release glucose for catabolism. Triglyceride is present already.

To understand these processes, we will review the following:

Glucose → Glycolysis (breakdown of glucose)
Triglyceride → Beta-Oxidation (breakdown of TG)
Amino acid → Gluconeogenesis (Conversion of a.a. → Kreb's cycle intermediates) Kreb's cycle (the citrus acid cycle → creation of reduced compounds for oxidation-reduction reactions → **ATP**

Oxidative Phosphorylation → **ATP**

With the creation of ATP, a high potential energy compound, the cell is in position to decompose ATP into energy to do all the cell's physiological processes. In the player, these processes are muscle, brain, cardiac, kidney function and sweating to cool down.

Carbohydrates, Fats, and Proteins → Glucose, Triglyceride, and Amino acid →

Glycolysis, Beta-Oxidation, and Gluconeogenesis →

Kreb's Cycle →

Oxidative Phosphorylation →

ATP →

Brain, Cardiac, Kidney, Muscle physiological processes = work.

Organic compounds = potential energy →

ATP = a higher energy compound →

energy → work.

Glycolysis

Complete combustion of glucose yields -2,840 kJ/mol. of standard free energy change.

Glucose + O_2 + Heat → CO_2 + H_2O + (-2,840 kJ/mol.)

Glucose = C, H, O atoms
O_2 = oxygen, the final electron acceptor
Heat = energy input to start and finish the process (endogonic)
endergonic = reaction requires an input of energy to proceed. The energy required to break the covalent bonds of glucose.
CO_2 = carbon oxidized.
Oxidation = O added to C.
H_2O = oxygen protonated.
Protonated = Reduction = H added to O.
Standard free energy change = negative = release of energy.
kJ = 1 thousand Joules of energy, unit of work.
mol. = amount of matter, a unit of weight.

exergonic = reaction releases energy.

Remember, energy can take the form of chemical bonds or heat.

A (-) $\Delta G'^o$ = STD free energy change means the reaction will proceed forward. There is enough energy for the reaction to proceed in the forward direction.

A (+) $\Delta G'^o$ means that the reaction requires energy to proceed forward.

$$\Delta$$
$$\text{Substrates = Reactants} \rightarrow \text{Products}$$

The overall equation for glycolysis is

Glucose + 2 NAD+ + 2 ADP + 2 Pi → 2 **Pyruvate** + 2 NADH + 2 H+ + 2 ATP + 2 H2O
6C   2 * 3C

NAD+ = oxidizing agent.(A vitamin B5 = niacin).

ATP = high energy compound. Energy is created in a high energy bond of phosphate. The potential energy is increasing.

Matter is conserved : 6C=2*3C.

H+ and electrons are in motion.

Oxidation-reduction

Oxidizing agent + compound → (compound + e) + (oxidizing agent + H+).
The oxidizing agent is reduced and the compound is oxidized.

Reducing agent + compound → (compound + H+) and (reducing agent + e).
The reducing agent is oxidized and the compound is reduced.

The overall equation can be divided into 2 separate equations.

Glucose + 2 NAD+ → 2 **Pyruvate** + 2 NADH + 2 H+ $\Delta G1'^o$ = -146 kJ/mol. Exergonic

2 ADP + 2 Pi → 2 ATP + 2 H2O $\Delta G2'^o$ = 61 kJ/mol. Endergonic

$\sum \Delta G'^o$ = -85 kJ/mol. Exergonic = proceeds to completion, releases energy = has enough energy to finish the job.

Glycolysis has 10 reaction steps and 2 phases: Preparatory and Payoff.

Step 1 is the first priming reaction which energizes glucose and traps G 6-P in the cytosol.

G 6-P has no protein transporters for it to leave the cytosol compartment. There is an input of energy to the reaction so that it may proceed forward.

Cyclic configurations are in the Hemiacetal form.

Step 1

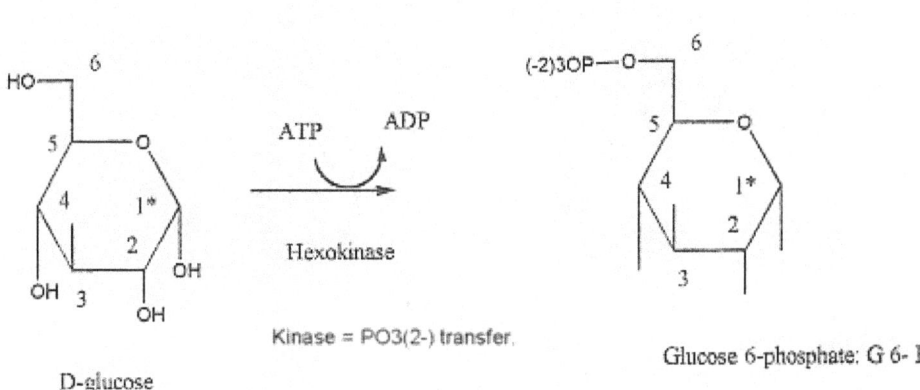

Glycolysis: C1*

ATP   ADP

Hexokinase

Kinase = PO3(2-) transfer.

D-glucose

Glucose 6-phosphate: G 6- P

* High energy bond

ATP

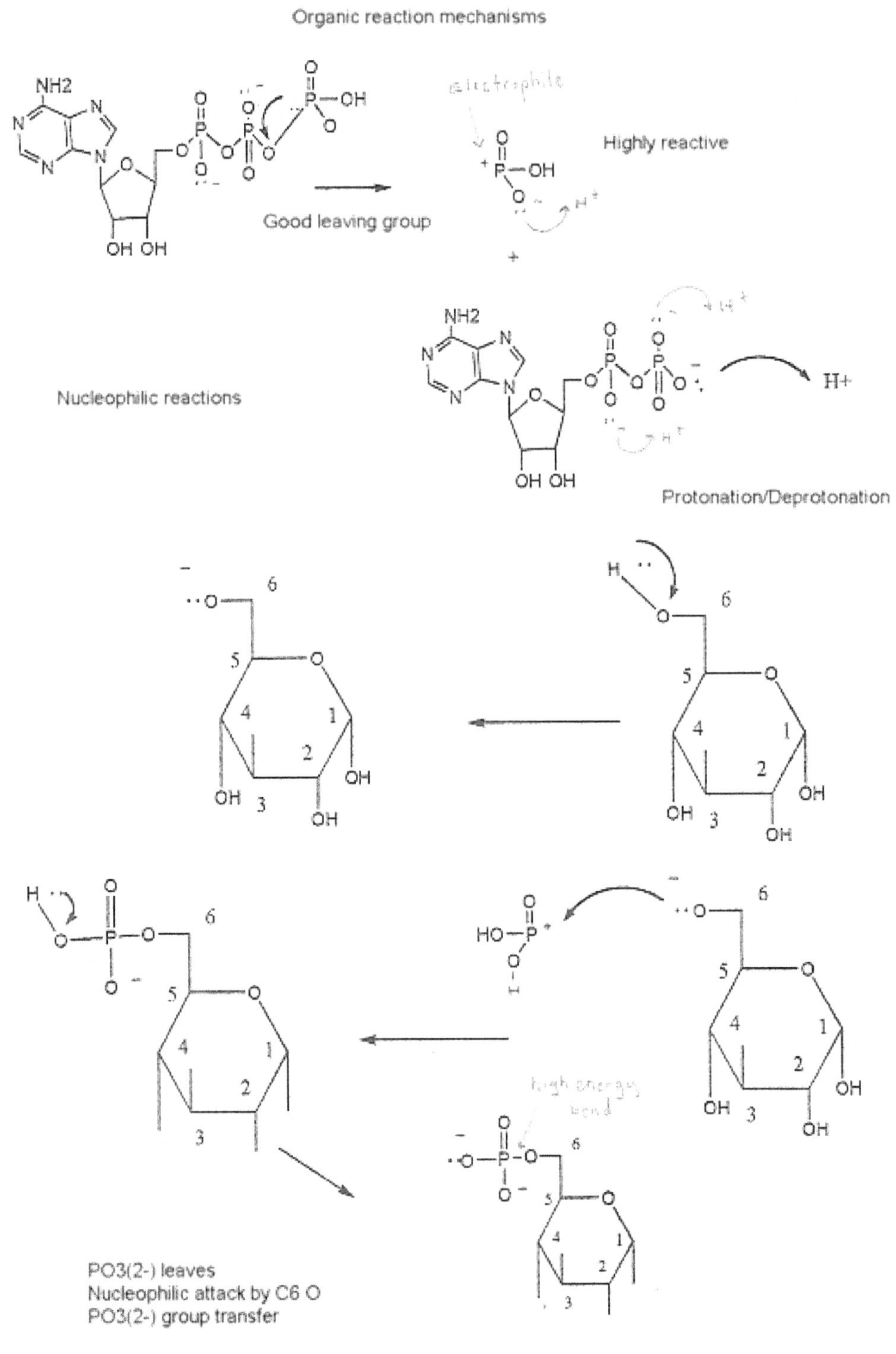

Organic reaction mechanisms

Highly reactive

Good leaving group

Nucleophilic reactions

Protonation/Deprotonation

H+

PO3(2-) leaves
Nucleophilic attack by C6 O
PO3(2-) group transfer

Step 2

Step 2

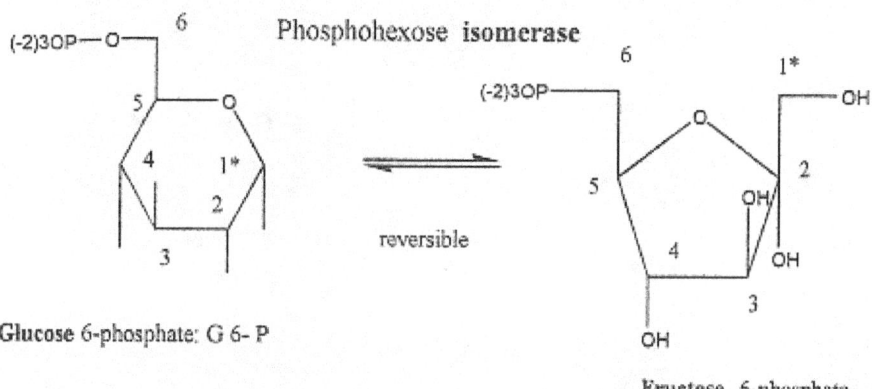

Phosphohexose isomerase

reversible

Glucose 6-phosphate: G 6- P

Fructose 6-phosphate

Oxonium ion

Isomer = same # of atoms just
different configuration

break and position

You can see as C 5 --> C2; C1* orients itself.

C#5 O stressed by enzyme electronically.

Acetal bond is stressed.
Alkoxide ion created.
C5 and C2 repositioned.
Nucleophilic attack of C5 O on C2 dipole.
Isomer = fructose

This seems like a very simple reaction mechanism. 1 bond broken and 1 bond formed. What does organic chemistry and the biologists say about this step 2. 1 bond is broken and 1 bond is formed. But there is acid-base chemistry with an unstable intermediate involved in the reaction. The unstable intermediate deprotonates the Glu residue of the enzyme eventually creating the ketone sugar. Finally, an intramolecular alcohol and carbonyl esterification takes place producing the isomer, fructose. The

His residue of the enzymes helps orchestrate the esterification process. As a point of interest, the enzyme is temporarily altered during the process.

Acetal is in the Hemiacetal form.

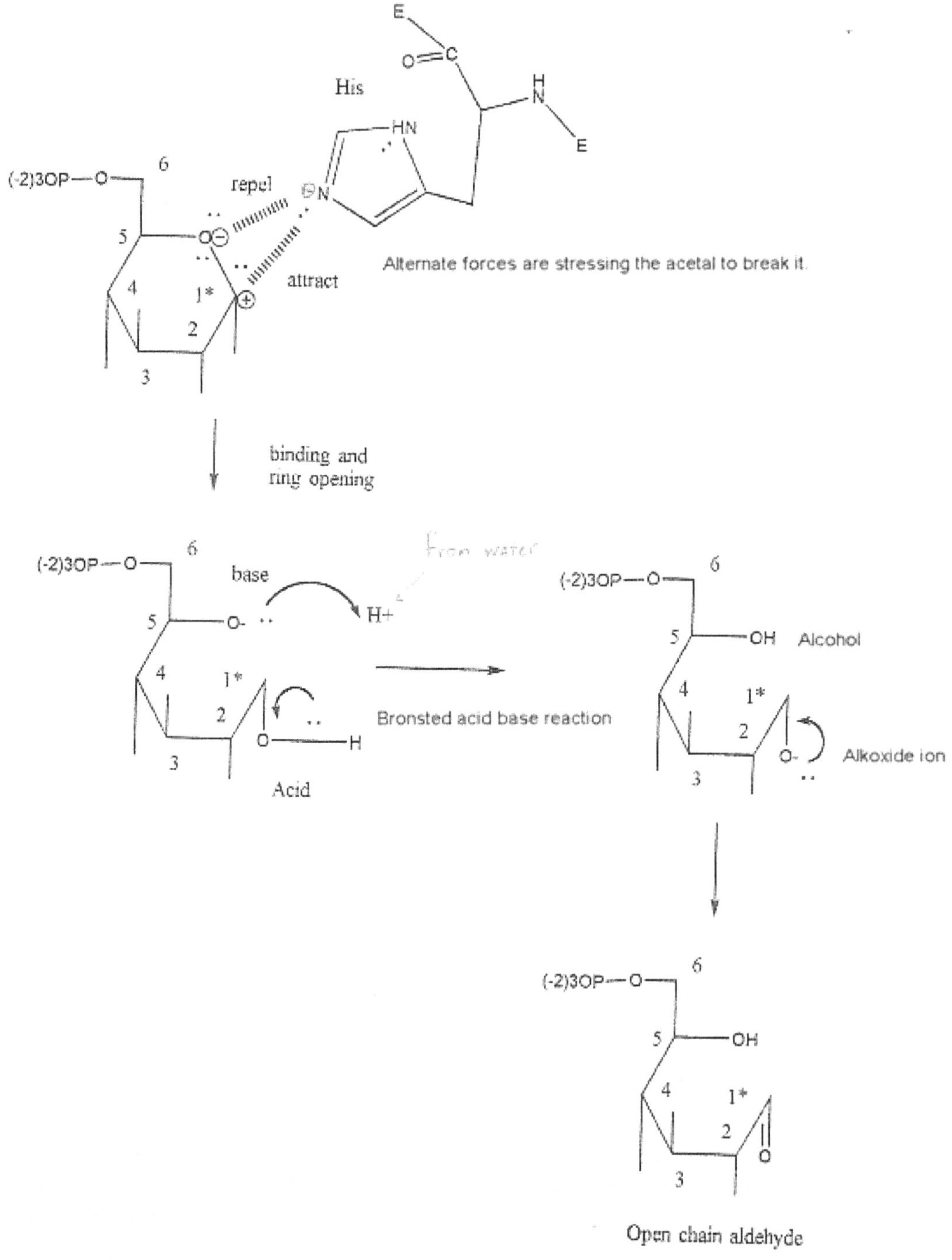

E
O=C
His
HN
N
E
(-2)3OP—O
6
5
4
1*
2
3
repel
attract
Alternate forces are stressing the acetal to break it.
binding and ring opening
base
O-
H+
From water
Acid
O—H
Bronsted acid base reaction
OH
Alcohol
O-
Alkoxide ion
O
Open chain aldehyde

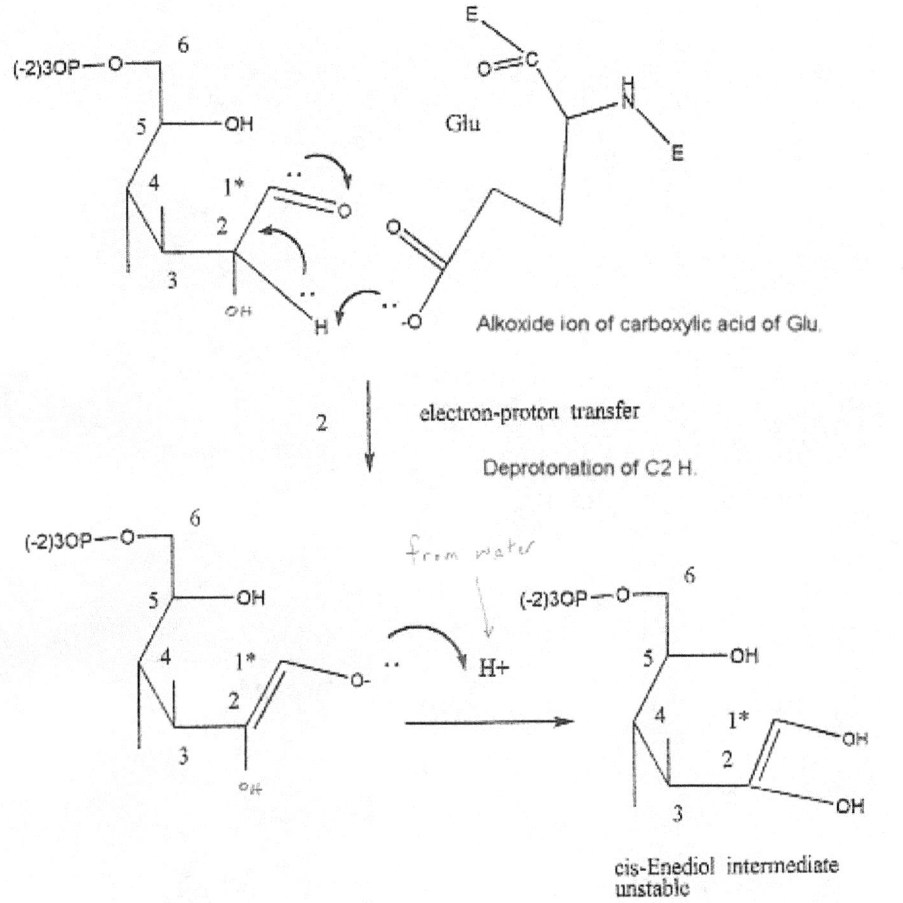
(-2)3OP—O
6
5
OH
4
1*
2
3
OH
H
O
E
C
H
N
E
Glu
-O
Alkoxide ion of carboxylic acid of Glu.
2
electron-proton transfer
Deprotonation of C2 H.
O-
from water
H+
OH
cis-Enediol intermediate
unstable

cis = OH on same side
ene = double bond
diol = double alcohol function

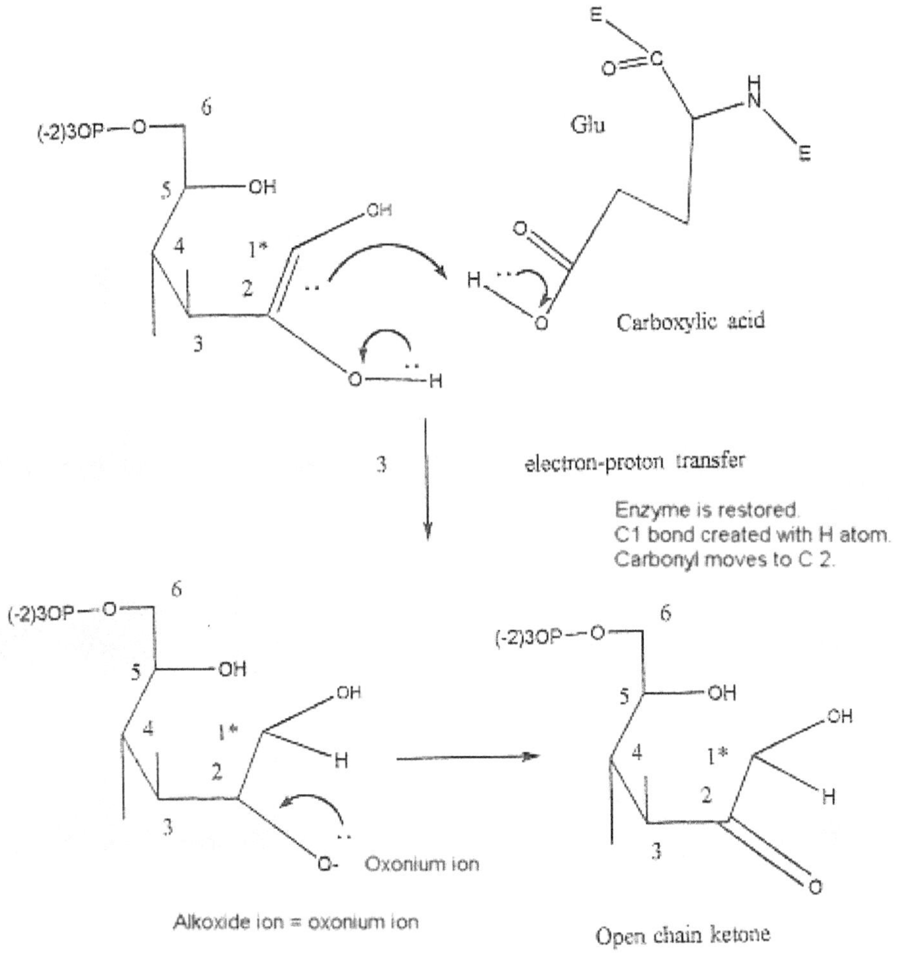

(-2)3OP—O
6
5
OH
4
1*
2
3
OH
O—H
Glu
E
O=C
H
N
E
O
H
O
Carboxylic acid
3
electron-proton transfer
Enzyme is restored.
C1 bond created with H atom.
Carbonyl moves to C 2.
H
O-
Oxonium ion
Alkoxide ion = oxonium ion
O
Open chain ketone

Push and pull of the His residue brings the H of the alcohol and oxygen of the carbonyl closer for esterification.

Acetal
closed ring
Internal alcohol and carbonyl esterification

Acetal is in the Hemiacetal form.

C1 – C2=O + HO-C5 → C1- C2-OH   Hemiacetal Carbonyl is protonated and Alkyl oxide is
   | | added to C2.
  C3   OC5

Step 3

Step 3   2nd priming reaction

Energy in : endergonic : irreversible

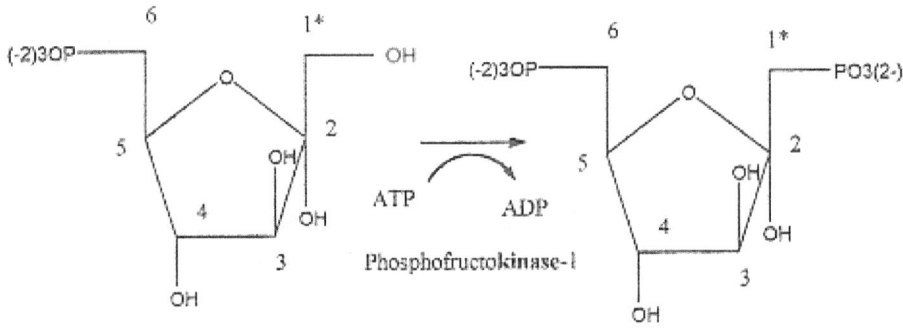

Fructose 6-phosphate

Fructose 1, 6-bisphosphate

Same organic reaction mechanism as step 1

The product is symmetrical somewhat.

Step 4

Step 4

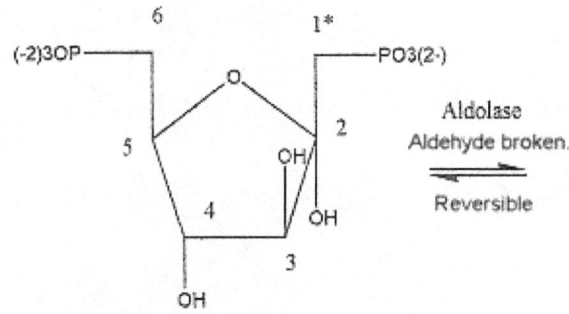

Aldolase
Aldehyde broken.

Reversible

Fructose 1, 6-bisphosphate

Acetal bond is broken.
Alkoxides are formed.
Bronsted Acid Base reactions.
Aqueous environment.

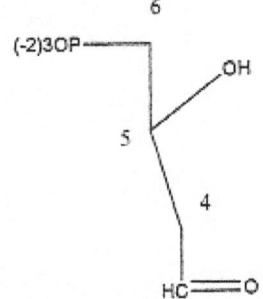

Glyceraldehyde 3-phosphate
G 3-P

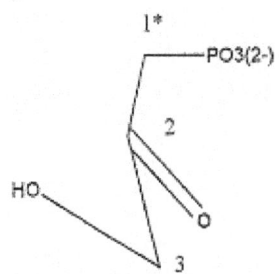

Dihydroxyacetone phosphate
DHAP
Ketone

Acetal = Hemiketal

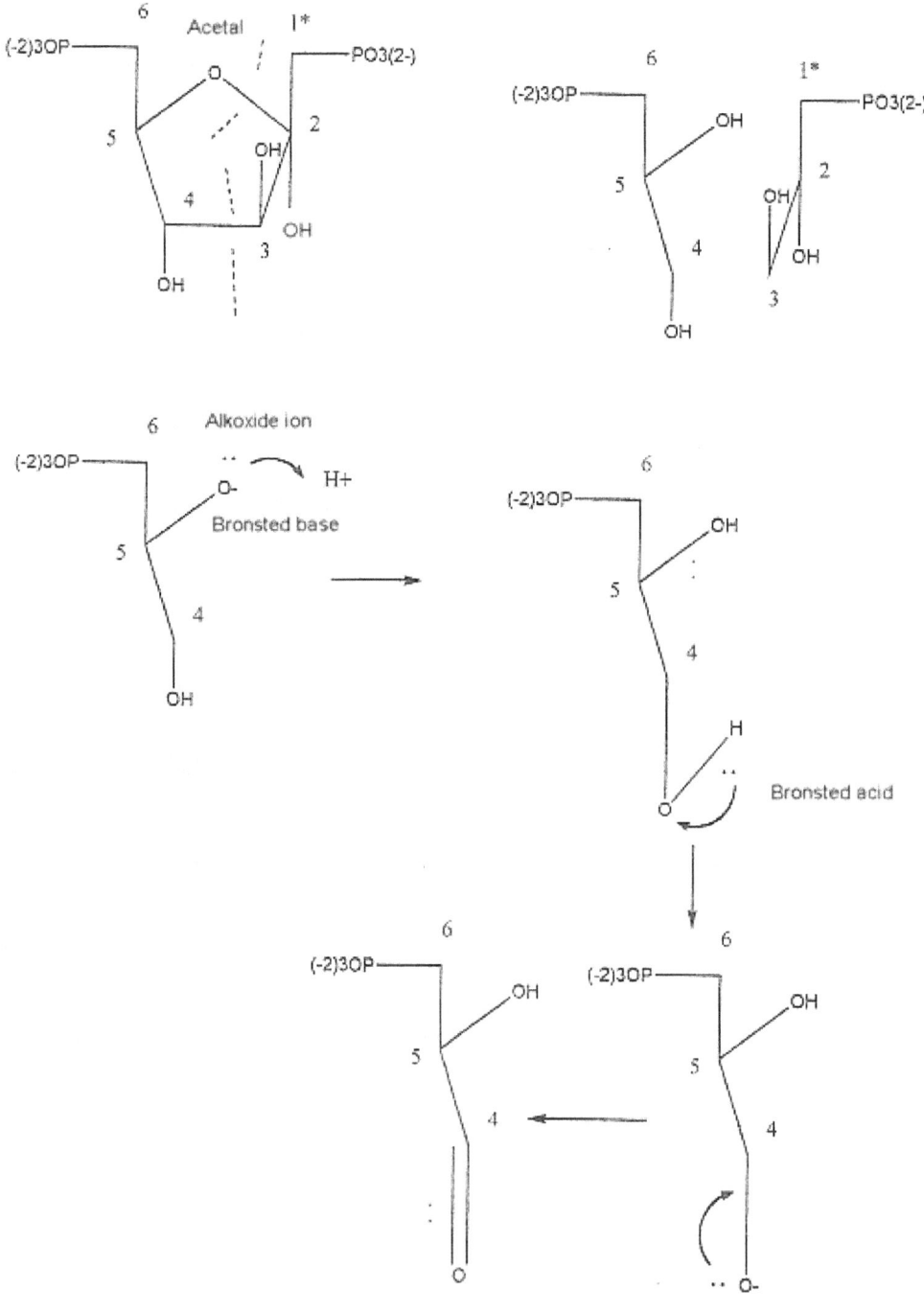

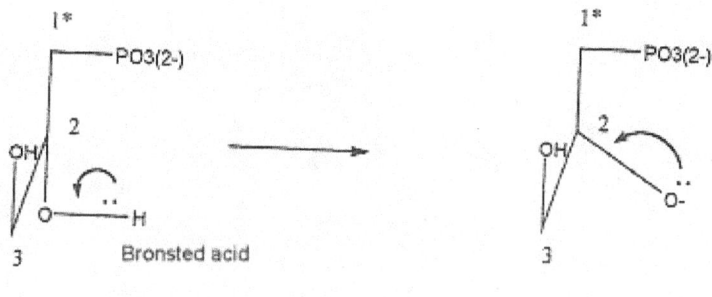

As noted above, there are a lot of acid base reactions. Intuitively, the above process seems plausible. But what do the biochemists think about this enzyme reaction. There is a multi-reaction sequence that takes place. In step 1, the hemiketal ring is opened by the nucleophilic function of the amino acid His. C2 is immediately protonated, and then deprotonated by water. C2 is now a carbonyl function. The Lys nucleophilic base creates a covalent bond with C2. This pushs the pi electrons to be protonated by the Asp acid function; thus creating an alcohol function at C2.

His stresses the hemiketal
Lys base forms covalent bond with C2
C2 carbonyl acidified by Asp Acid to become alcohol

Even though the arrow is unidirectional, it is really reversible also.

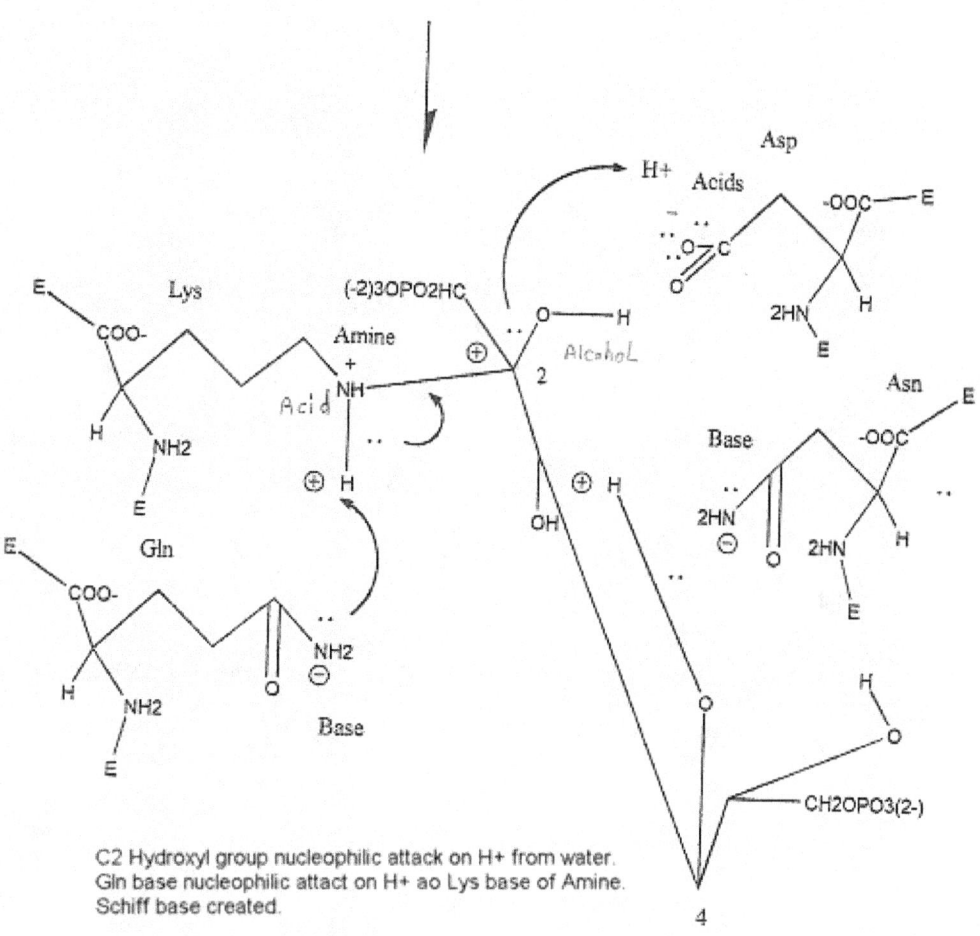

C2 Hydroxyl group nucleophilic attack on H+ from water.
Gln base nucleophilic attact on H+ ao Lys base of Amine.
Schiff base created.

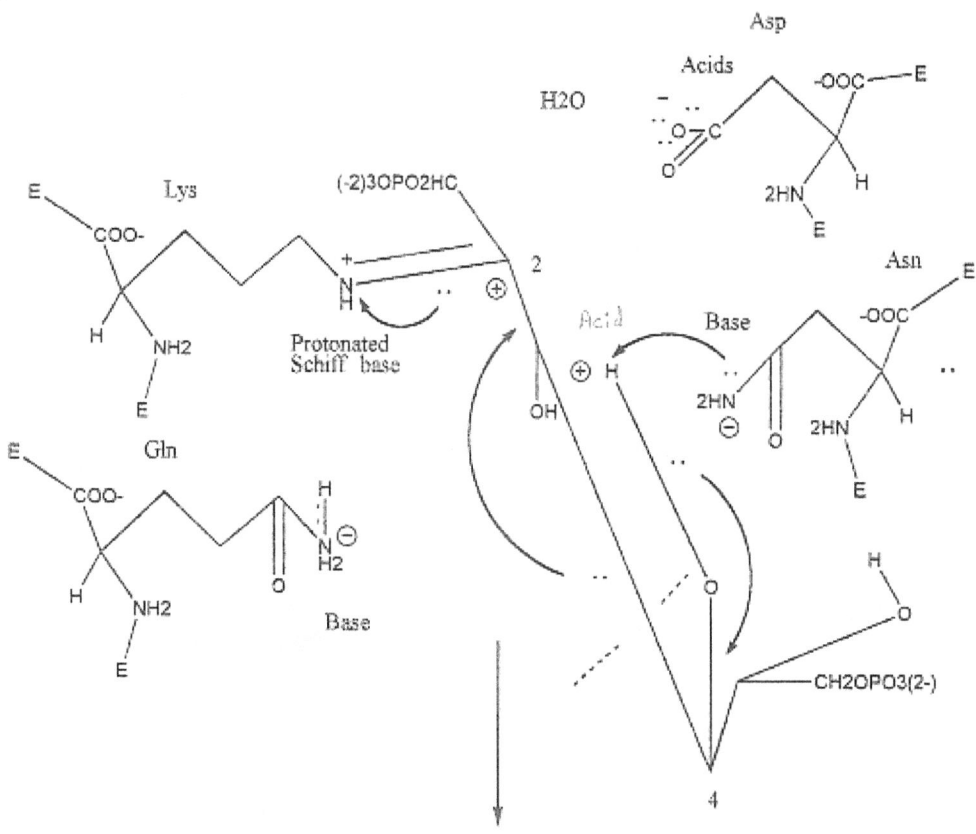

Asn base nucleophilic attack of C4 alcohol H+.
Covalent bond between C3-4 broken due to electron couple shift.
Shift occurs to Schiff base.
Enamime intermediate created.
En=double bond
amine=Nitrogen
G 3-P created.

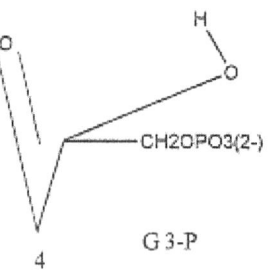

G 3-P

+

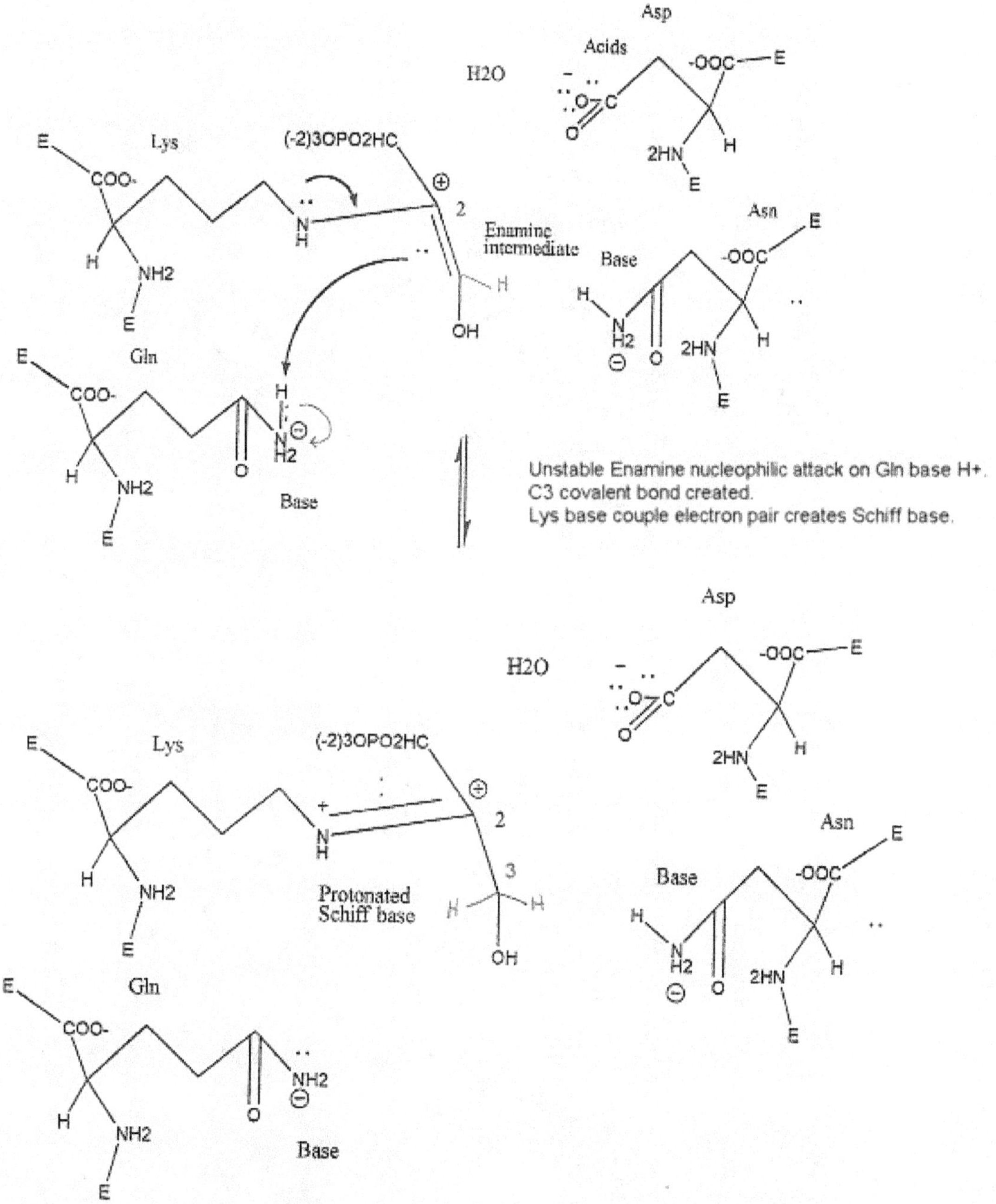

Asp

Acids

H2O

-OOC—E

:O::C

O::C

2HN

H

E

E
Lys

COO-

H

NH2

E

(-2)3OPO2HC

⊕

N
H

2

Enamine
intermediate

H

OH

Asn

E

Base

-OOC

H
N
H2
⊖

O

2HN

H

E

E
Gln

COO-

H

NH2

E

H

N ⊖
H2

O

Base

Unstable Enamine nucleophilic attack on Gln base H+.
C3 covalent bond created.
Lys base couple electron pair creates Schiff base.

Asp

H2O

-OOC—E

:O::C

O::C

2HN

H

E

E
Lys

COO-

H

NH2

E

(-2)3OPO2HC

⊕

+
N
H

2

Protonated
Schiff base

3

H

H

OH

Asn

E

Base

-OOC

H
N
H2
⊖

O

2HN

H

E

E
Gln

COO-

H

NH2

E

NH2
⊖

O

Base

Hydrolysis, water, reverses the Schiff Base.
Water Hydroxyl group nucleophilic attack of C2.
Pi electrons jump to base of Lys.
Carboxylic acid function of Asp nucleophilic attack of C2 H+ of alcohol.
Base of Lys nucleophilic attack of H+ of water.
C2 carbonyl created.

86

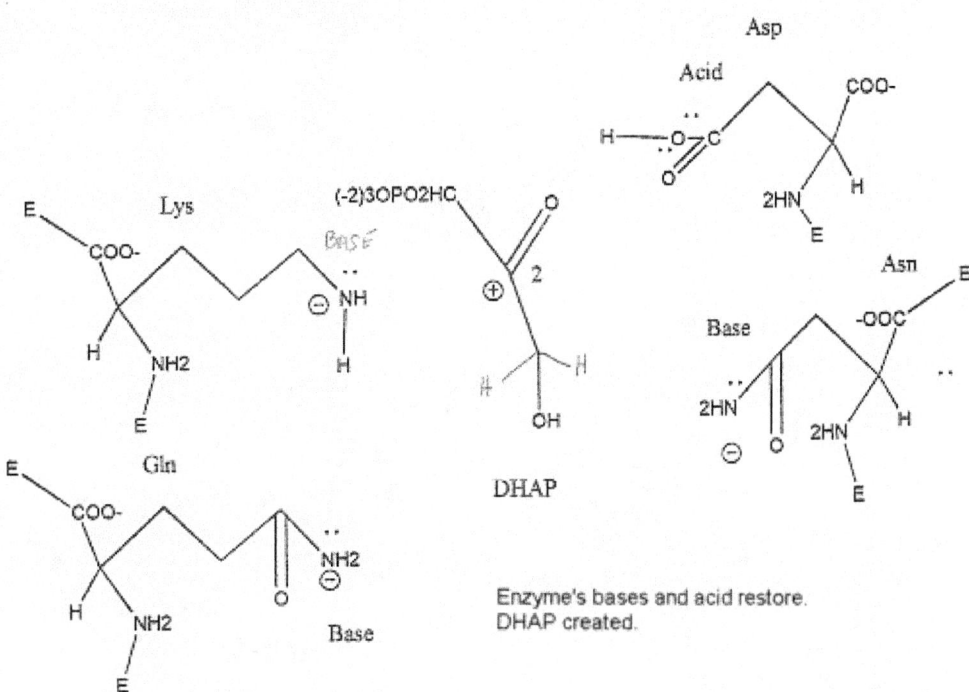

DHAP

Enzyme's bases and acid restore.
DHAP created.

Step 5

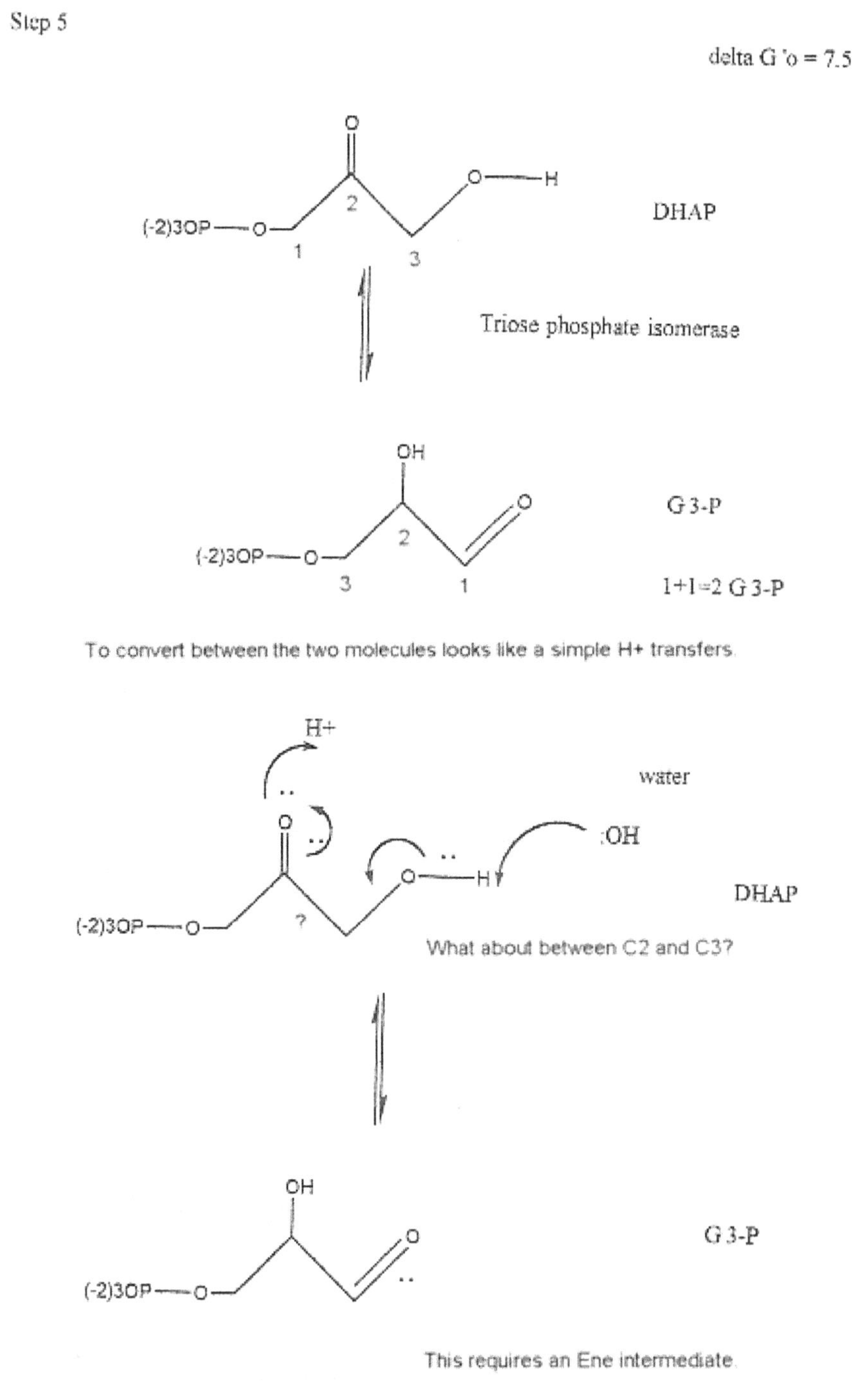

Step 5

delta G 'o = 7.5

DHAP

Triose phosphate isomerase

G 3-P

1+1=2 G 3-P

To convert between the two molecules looks like a simple H+ transfers.

H+

water

:OH

DHAP

What about between C2 and C3?

OH

G 3-P

This requires an Ene intermediate.

There is a small change in energy for the conversion process.

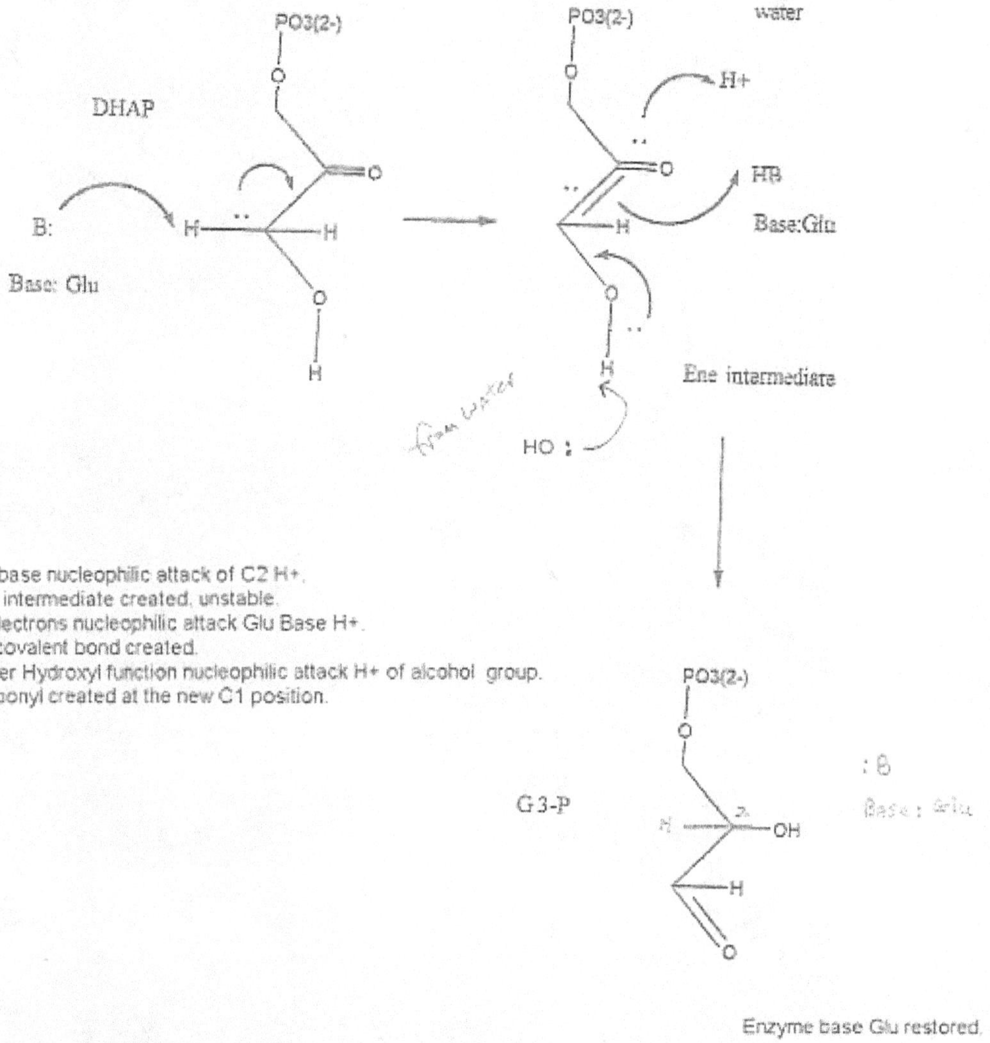

PO3(2-)

DHAP

B:

Base: Glu

from water

PO3(2-)

water

H+

HB

Base:Glu

Ene intermediate

HO :

Glu base nucleophilic attack of C2 H+.
Ene intermediate created, unstable.
Pi electrons nucleophilic attack Glu Base H+.
C2 covalent bond created.
Water Hydroxyl function nucleophilic attack H+ of alcohol group.
Carbonyl created at the new C1 position.

G 3-P

PO3(2-)

OH

H

:B

Base : Glu

Enzyme base Glu restored.

B: is short hand for nucleophilic base like Glutamate.
Glutamate is the base. Glutamic acid is the protonated base.
Can you draw the base. Hint = Carboxylic acid minus the H+.

Step 6

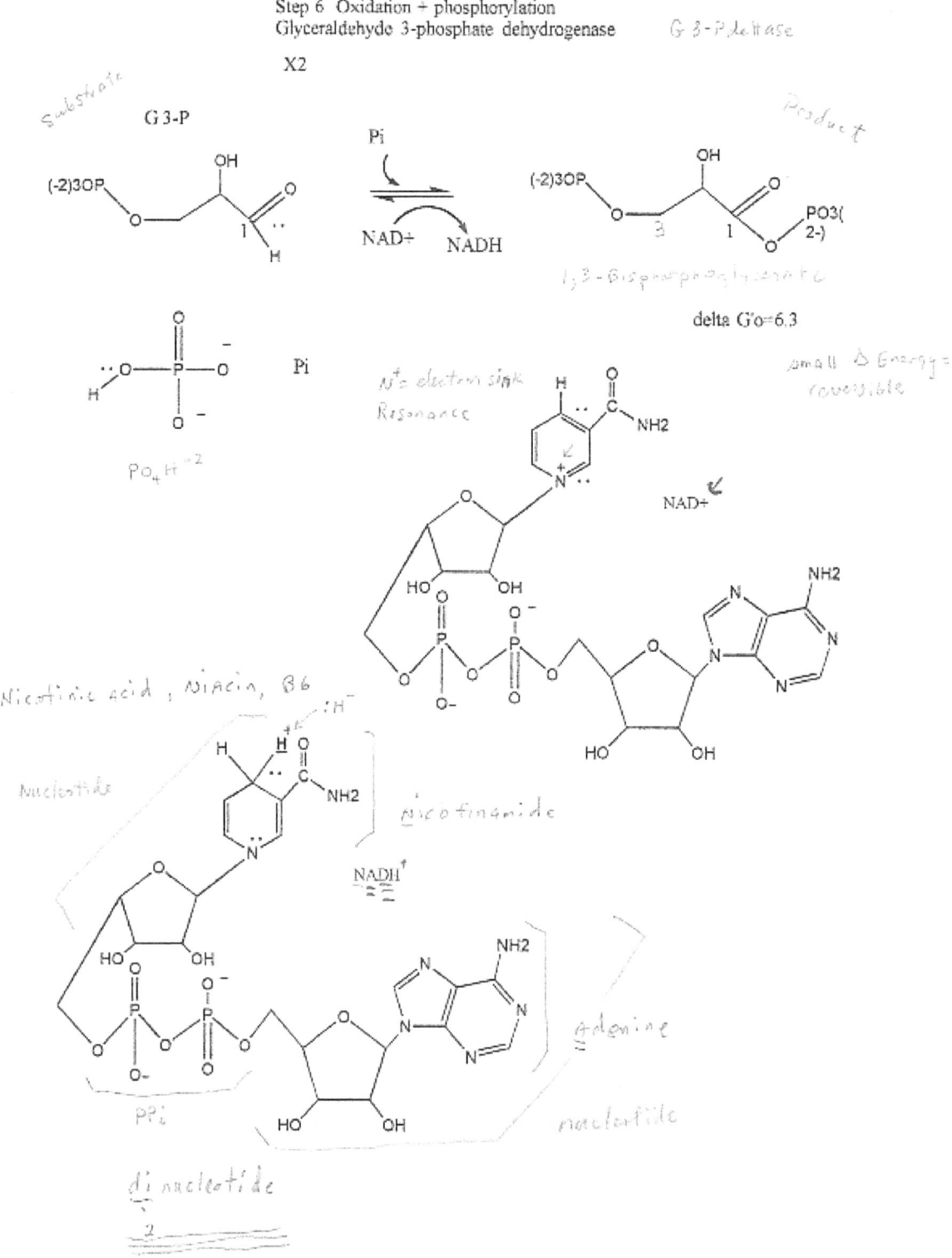

Step 6 Oxidation + phosphorylation
Glyceraldehyde 3-phosphate dehydrogenase   G 3-P dehase

X2

Substrate   G 3-P   Product

Pi

NAD+   NADH

1,3-Bisphosphoglycerate

delta G'o=6.3

Pi

PO₄H⁻²

N's electron sink
Resonance

small ∆ Energy =
reversible

NAD+

Nicotinic acid , Niacin, B6

Nucleotide

nicotinamide

NADH

Adenine

PPi

nucleotide

di nucleotide

E=enzyme

The reaction pocket of the enzyme has 2 covalent residues : Cys + His. And 1 loosely bound oxidizing agent : NAD+. The 2 residues are Hydrogen bonded by base of His and acid of Cys.

Formation of enzyme-substrate complex

His base nucleophilic attack H+ of Cys.
Sigma bond electron pair nucleophilic attack C1 of carbonyl.
Pi bond electron pair creates oxonium ion.

Note the sigma electrons of the Cys covalent bond attacking the C1.

Formation of thiohemiacetal intermediate

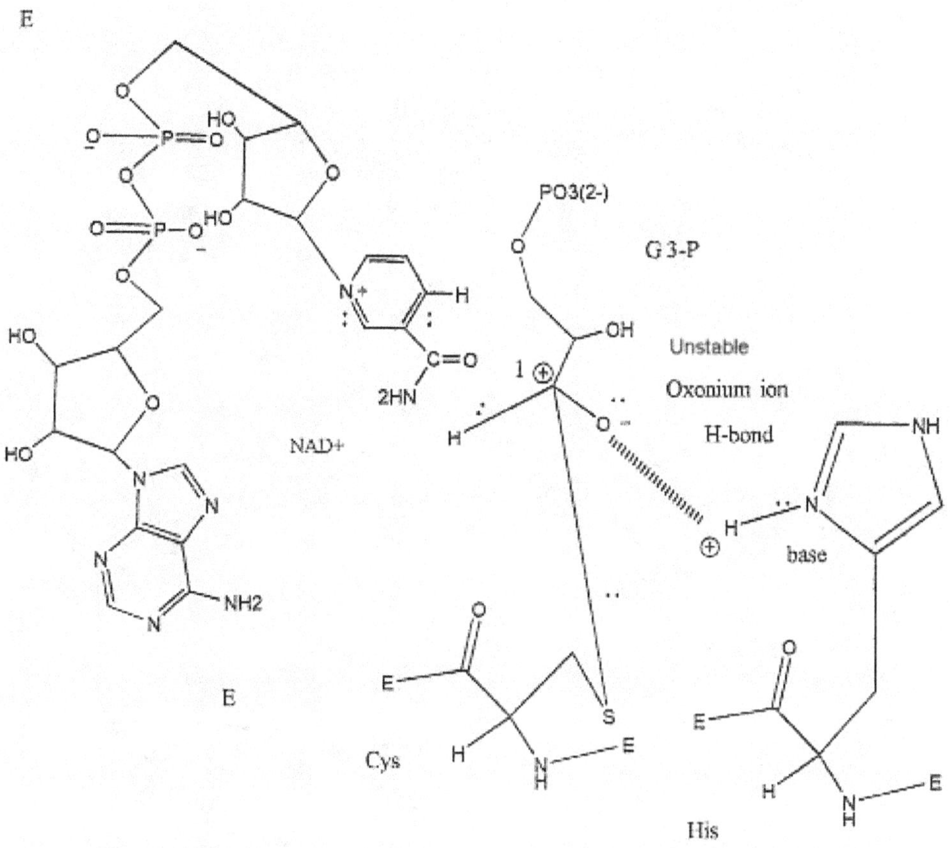

E

HO

G 3-P

$PO_3(2-)$

Unstable

Oxonium ion

H-bond

NAD+

base

Cys

His

The oxonium ion is unstable. It usually reverts back to carbonyl function.
Thio = S
Hemi = ½
Acetal = R-O + R-S on C1.

Oxidation to thioester intermediate

Oxidation means you lose electrons. G 3-P loses a Hydride because of NAD+ sink.

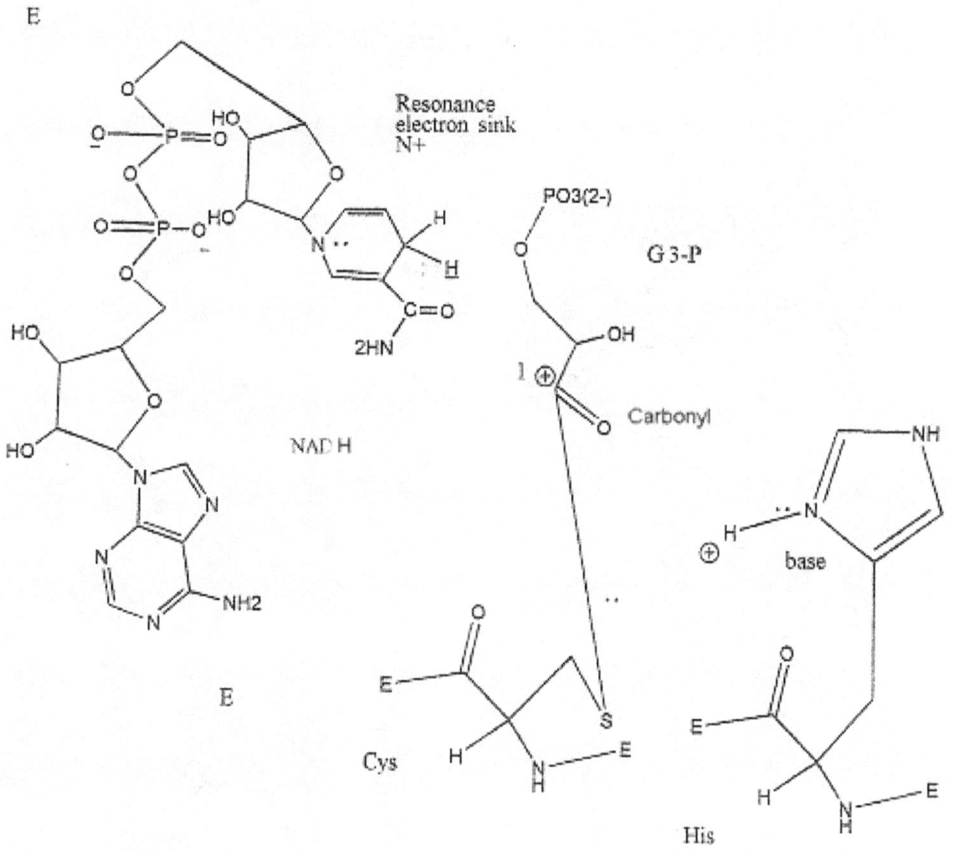

E

Resonance
electron sink
N+

PO3(2-)

G 3-P

HO

HO

HO

O—P=O

O=P—O

HO

HO

H

N

H

H

C=O

2HN

NADH

OH

Carbonyl

1 ⊕

O

NH

N

H

base

⊕ H

NH2

N

N

E

O

E

S

Cys

H

N
H

E

O

E

His

H

N
H

E

Note the N+ lost it's positive charge in NAD+.

NADH will leave and be replaced by NAD+. The reduced compound will proceed to oxidative phosphorylation to give up it's Hydride electrons to make ATP (energy).

Sigma electron pair attacking the protonated His base.
Restores the enzyme to original state.
Product released : 1,3-Bisphosphoglycerate (high energy compound).

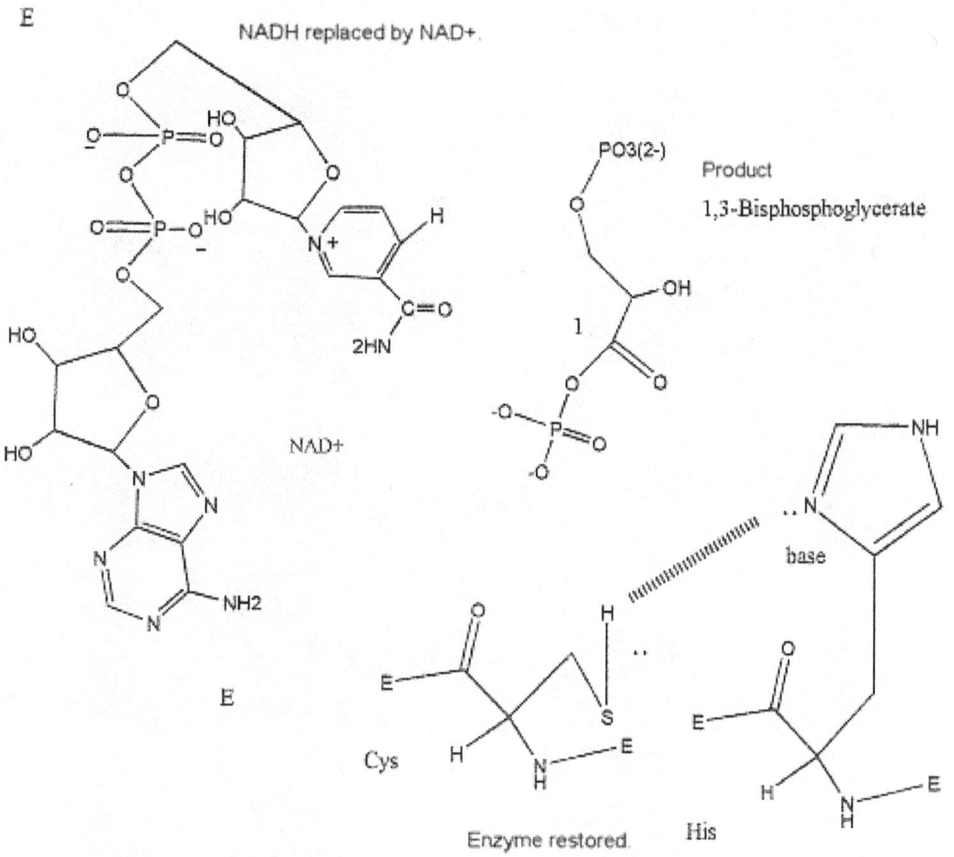

NADH replaced by NAD+.

NAD+

PO3(2-)

Product

1,3-Bisphosphoglycerate

base

Cys

His

Enzyme restored.

Niacin, B6, plays an important role in the energy production during carbohydrate degradation. Note the enzyme is unloaded once again for another round.

Step 7

Step 7 1,3-Bisphosphoglycerate --> 3-Phosphoglycerate   Phosphoglycerate kinase
2 ATP = substrate-level phosphorylation
Mg2+
delta G'o = -18.5
irreversible
X2

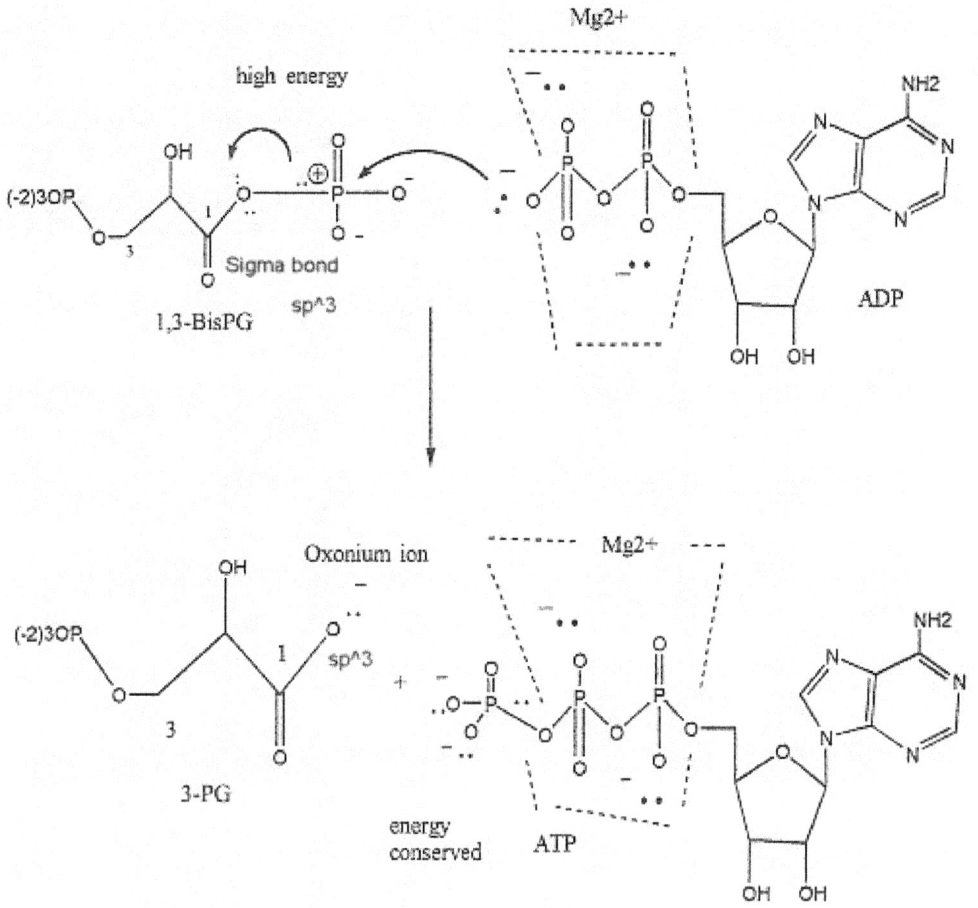

high energy

Mg2+

NH2

(-2)3OP

1

Sigma bond

sp^3

1,3-BisPG

ADP

OH

OH

Oxonium ion

Mg2+

NH2

(-2)3OP

OH

1

sp^3

3

3-PG

+

energy
conserved

ATP

OH

OH

Step 8

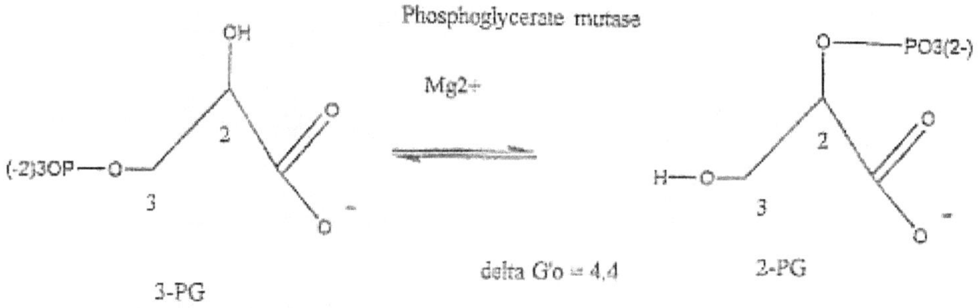
Phosphoglycerate mutase
Mg2+
OH
2
3
(-2)3OP—O
O
O
3-PG
delta G'o = 4.4
O—PO3(2-)
H—O
2-PG

The phosphate source for His is 2,3-BPG which is re-cycled.

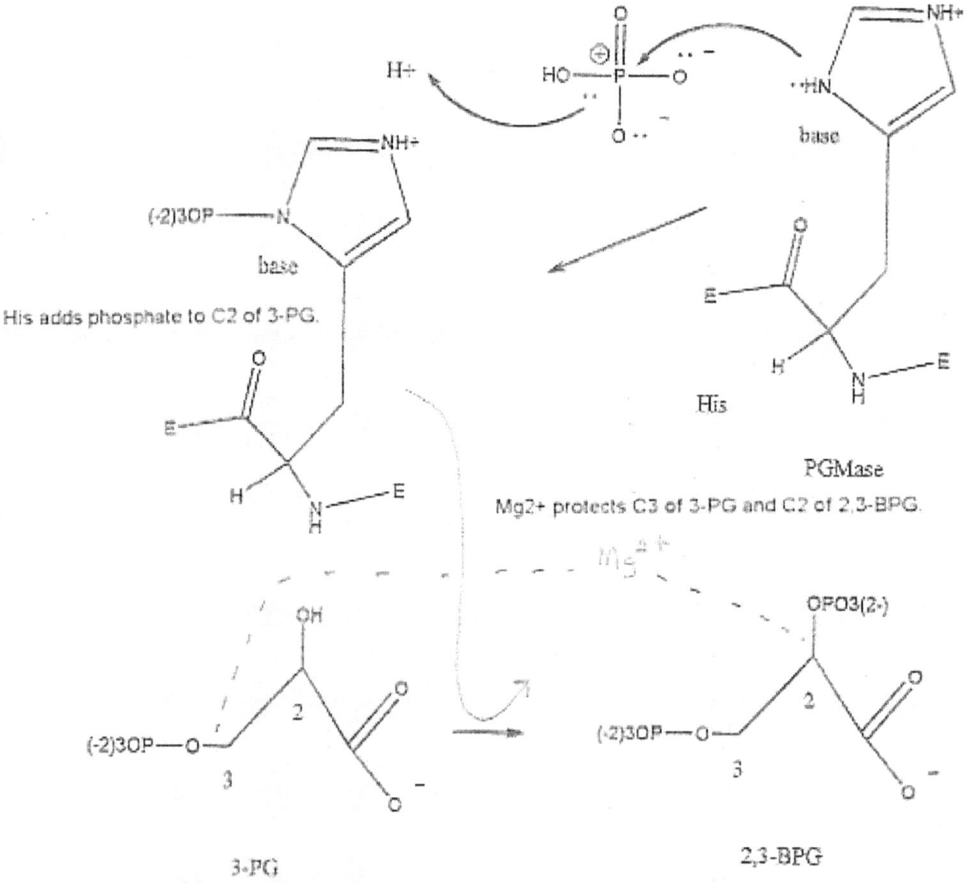
H+
HO
P
O
NH+
HN
base
(-2)3OP
N
His adds phosphate to C2 of 3-PG.
E
H
N
H
His
PGMase
Mg2+ protects C3 of 3-PG and C2 of 2,3-BPG.
OH
(-2)3OP—O
3-PG
OPO3(2-)
2,3-BPG

The + charge resonates between both N in His. But it is not really the + charge, it is the lone electron. The positive field expands from 1 N to the other N atom.

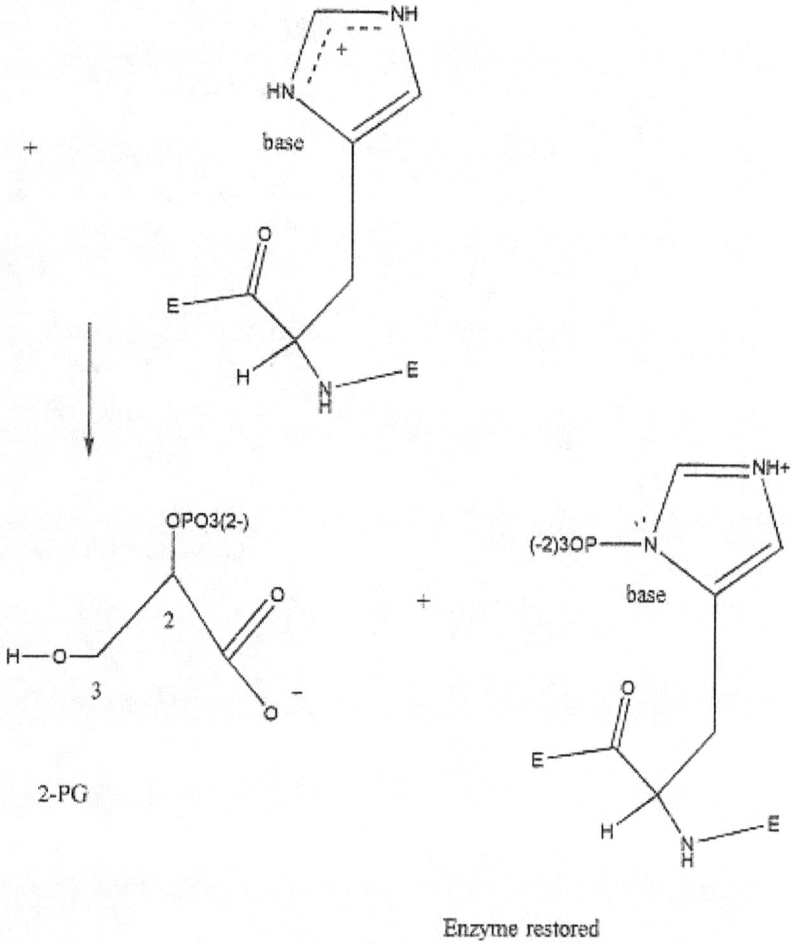

base

+

2-PG

OPO3(2-)

H—O

2

3

+

base

(-2)3OP

NH+

Enzyme restored

The Mg2+ protects the C positions of the substrate and intermediate to create the product, 2-PG.

Step 9

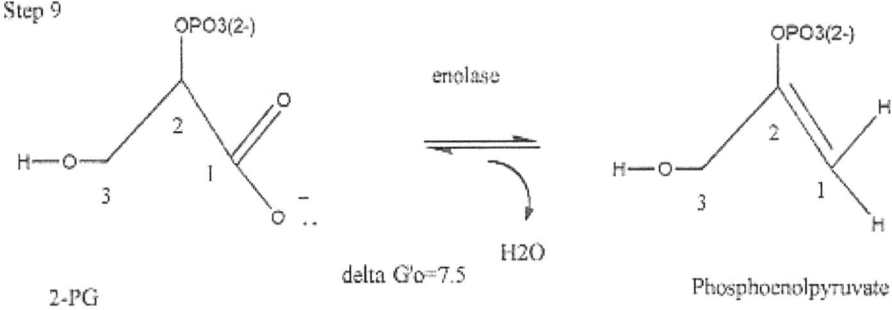

Step 9

2-PG

enolase

delta G'o=7.5

H2O

Phosphoenolpyruvate

2 Mg2+

Coordinate bonds : protects O atoms from the reaction.

2-PG

base : nucleophile

lone electron pair

E—NH

NH2

H

E

O

Lys 345 = a.a. residue #

Acid : Electrophile

H+

E—NH

O

H

E

O

O—H

Glu 211

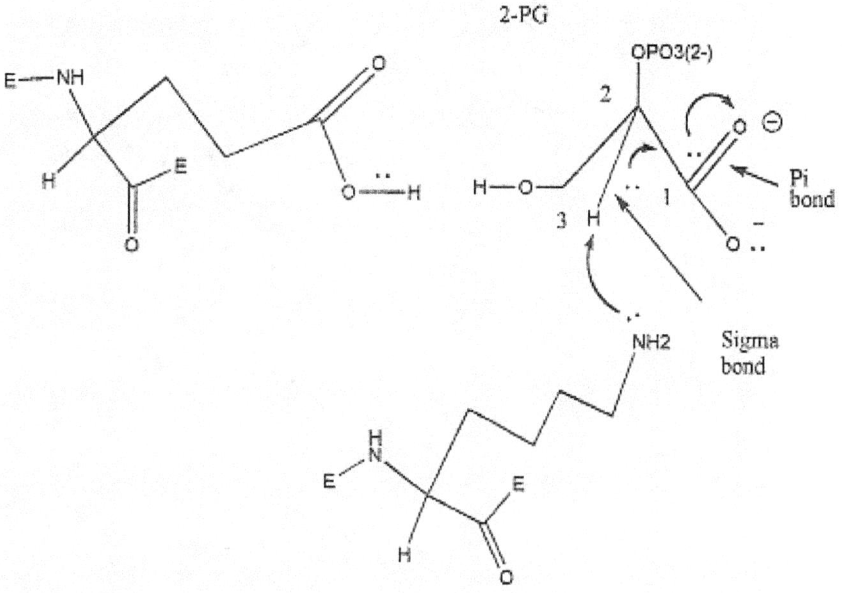

2-PG

E—NH

OPO3(2-)

2

H—O

3 H 1

Pi
bond

Sigma
bond

NH2

E—N
H

E

H

O

enolic intermediate
unstable

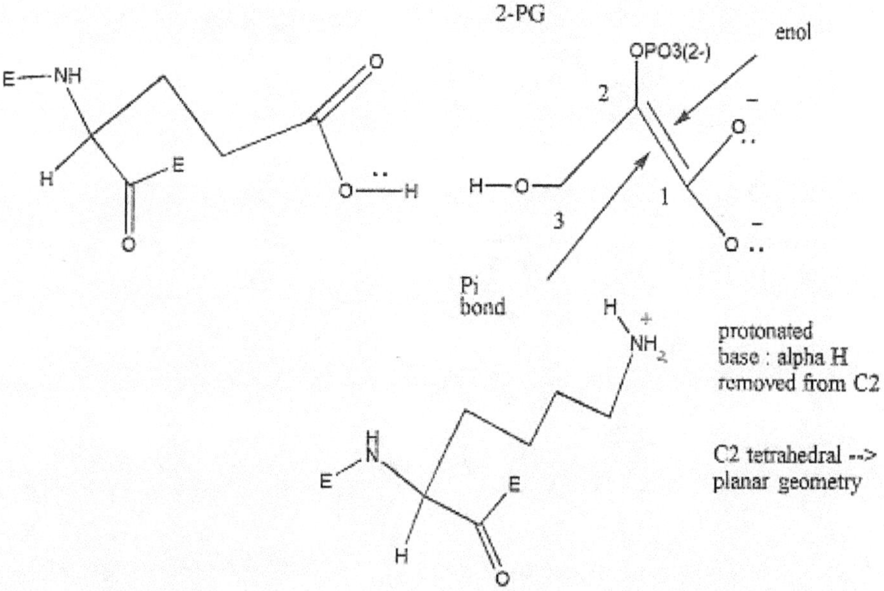

2-PG

enol

E—NH

OPO3(2-)

2

H—O

3 1

O

O

Pi
bond

NH2

H

protonated
base : alpha H
removed from C2

C2 tetrahedral -->
planar geometry

E—N
H

E

H

O

2-PG

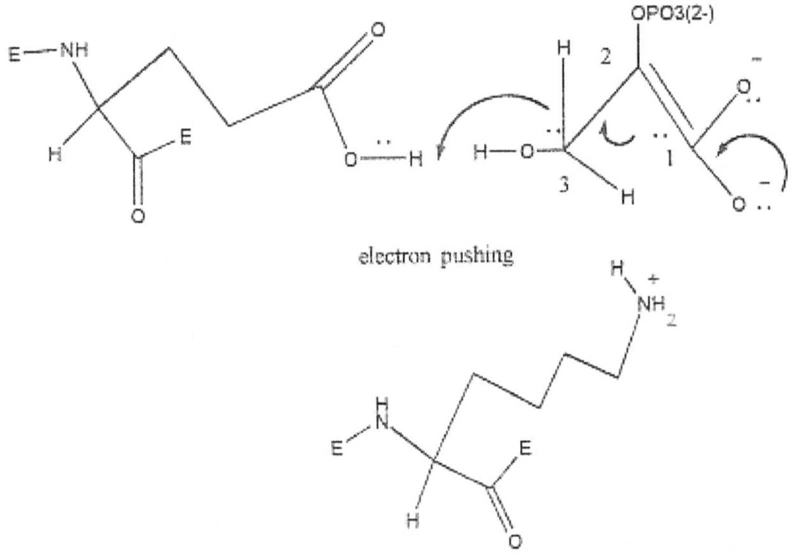
OPO3(2-)
electron pushing

Phosphoenolpyruvate

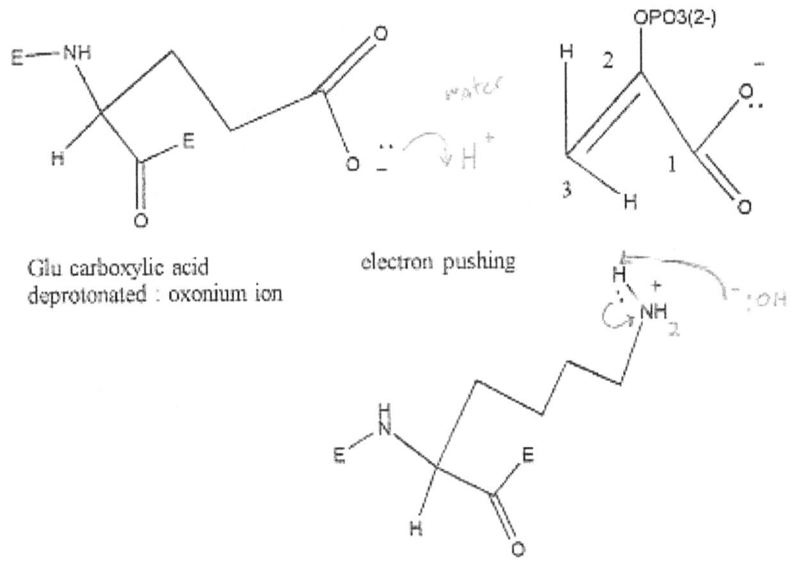
OPO3(2-)
water
electron pushing
Glu carboxylic acid
deprotonated : oxonium ion

+ H_2O

dehydration

Step 10

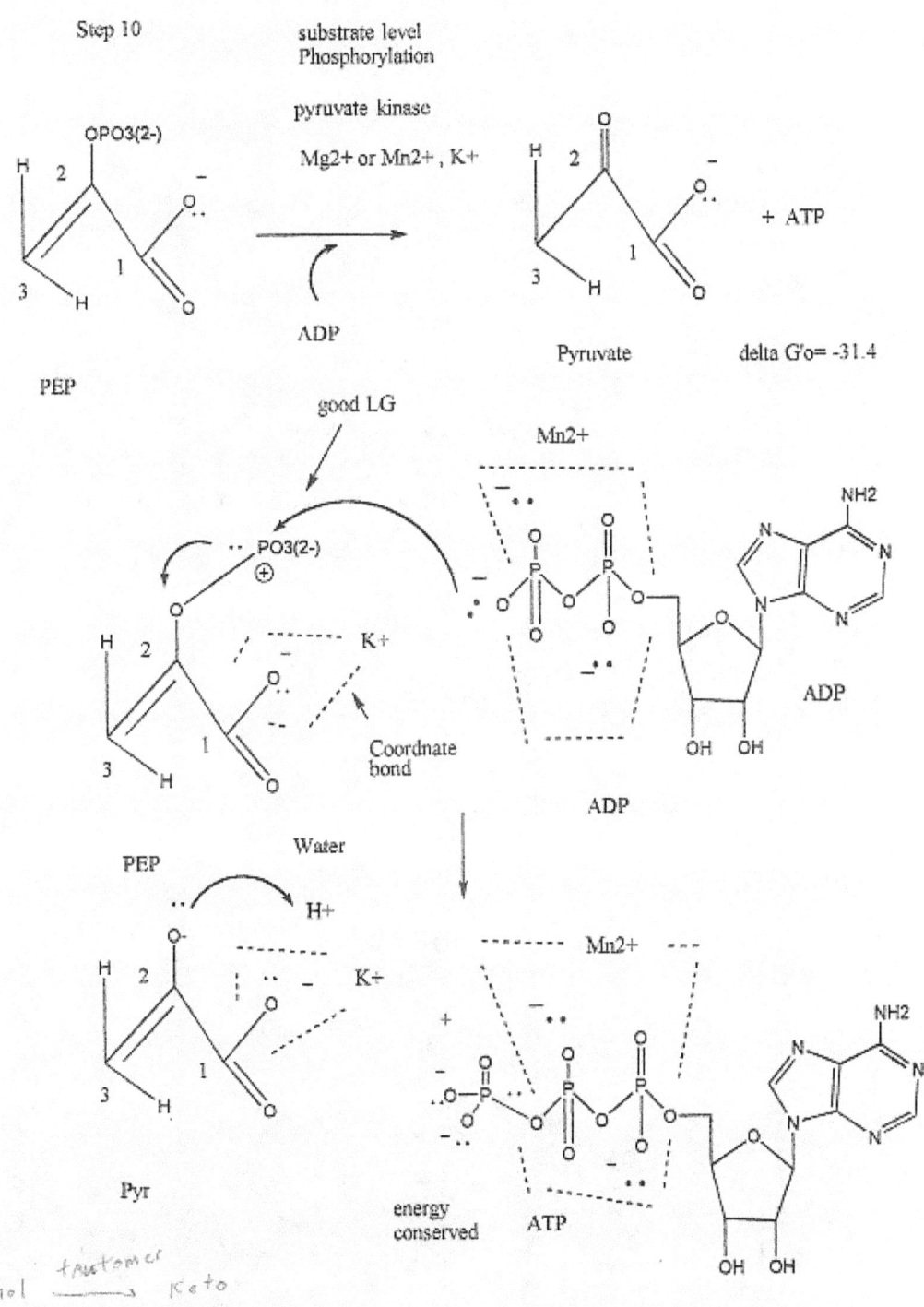

The overall equation for glycolysis

Glucose
+
2 NAD+
+
2 ADP
+
2 Pi

$\rightarrow$

2 pyruvate $\rightarrow$ Kreb's cycle
+
*2 NADH $\rightarrow$ Oxiidative phosphorylation $\rightarrow$ ATP
+
2 H+
+
*2 <u>ATP</u>
+
2 H2O

Thermodynamic Balance Sheet

Step	$\Delta G'^{\circ}$ (kJ/mol)
1	-16.7
2	1.7
3	-14
4	23.8
5	7.5
6	6.3
7	-18.5
8	4.4
9	7.5
10	-31.4
Net	-29.4 exergonic = force driving reaction forward

Chapter 8

Energy   2

From chapter 5, our product of lipid digestion was 2 fatty acids and 1 2-monoglyceride. These products need to be packaged for transport through the blood to their final destination. Both products are from the even numbered triglyceride, Tristearin = C18. As you will see, the even carbon fatty acids are handled differently than odd numbered acids. Also there is a difference in handling with higher numbered versus low numbered fatty acids. In our case, the tristearin is oxidized by a highly complex-ed enzyme initially; then finished by cytosolic, soluble enzymes. In our situation of a football player, the fat will be oxidized for energy either initially or at least, early in the process of the game. So let us proceed with following the fate of tristearin in our football player. The following outline should help:

      1. Packaging and transport
      2. Oxidation → energy

Packaging and Transport

There are 7 steps needed to get the TG to it's final destination.

1. The solubilization phase starts in the duodenum with **bile salts** from the Gall Bladder. These salts are polarized which makes them water soluble. They associate with the products of digestion and create micelles which act like water soluble ferries for absorption of the product. The emulsification process acts like a washer machine to break the digestants up into smaller sizes.
2. **Intestinal lipase** digests the TG into smaller products, eg. f.a..
3. Diffusion occurs through the intestinal mucosa and the products are reassembled into TG again. So Tristearin could be re-created, but most likely it is converted into another type of TG.
4. The TG is packaged into **chylomicrons** which are made of cholesterol, proteins and TG. The proteins are apolipoproteins which act as receptor hookups for the final destination.
5. **Apolipoprotein C-II** is destined for muscle and adipose tissue.
6. **Lipoprotein lipase** in the capillary blood in the tissue breaks TG down into glycerol and f.a..
7. Muscle tissue oxidizes f.a. for energy.
   Adipose tissue re-esterifies f.a. for storage.

As a side note, In the liver, TG can be used for energy or ketone synthesis.
With excess f.a., the TG as VLDL is transported to Adipose tissue to be used for storage.

Lipid processing

Solubilization

1. Gall bladder --> Bile salts --> small intestine emulsification

2. Small intestine --> intestinal lipase
   Tristearin --> f.a., mono, diacylglycerol, and glycerol

3. Diffusion through intestinal mucosa --> triacylglyceride

4. Packaging with cholesterol + proteins --> apolipoproteins=receptor -->
chylomicrons

5. Apolipoprotein C-II --> muscle + adipose cells

6. Lipoprotein lipase TG --> glycerol + f.a.

7. Muscle --> f.a. oxidized for energy
   Adipose --> f.a. reesterified for storage

T.G. in the liver --> energy or ketones

Excess f.a. --> T.G. as VLDL --> Adipose for storage

So for our football player, the possible TG, triglyceride, pathway could be as follows:

1. Breakfast or dinner meal of 15% fat. Most of it TG.
2. Bile salt and lipase digestion of TG into f.a. and glycerol.
3. Diffusion and absorption with resynthesis of f.a. and glycerol into TG.
4. Packaging into chylomicron with protein type Apolipoprotein C-II.
5. Lipase digesting TG into glycerol and f.a. again.
6. Muscle oxidizing f.a. into energy and glycerol oxidized by the Kreb's cycle for energy.

After the action of lipoprotein lipase, glycerol and f.a. are available for further oxidation or catabolism.
Glycerol is converted to a glycolytic intermediate to be oxidized in the glycolysis pathway.
F.a. is further oxidized in Beta-oxidation in the mitochondria.
Before this can happen and because our f.a. is > than 12 C, the f.a. takes the carnitine shuttle to enter the mitochondria.

In the outer mitochondrial membrane cytosolic side:
f acyl-CoA can either be oxidized or synthesized to membrane lipids.
1.acyl-CoA synthetases
F.a.+CoA+ATP-->f acyl-CoA+AMP+PPi delta Go'= -34 kJ/mol
2. carnitine acytransferase I
passage via a protein porin to the intermembrane space
3. acyl-carnitine/carnitine transporter : inner mitochondrial membrane
4. carnitine acyltranferase **II**
f.acyl group transferred from carnitine to intramitochondrial CoA.

We will further develop these steps: First the fate of glycerol

glycerol kinase

Pi transfered to glycerol
energizes the molecule

Glycerol
3C Alcohol

L= to the Left

L-Glycerol 3-P

ATP

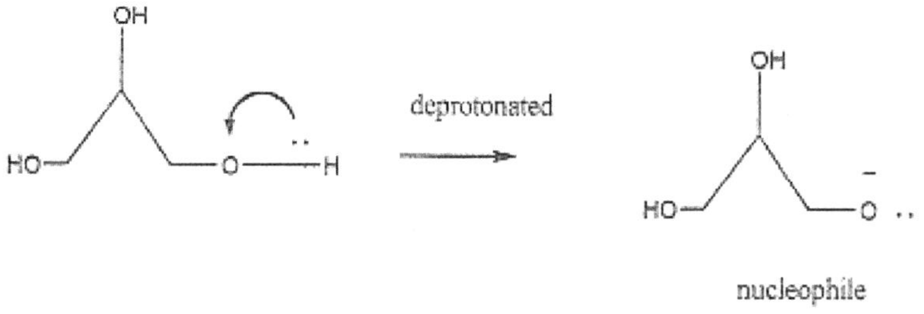

deprotonated

nucleophile

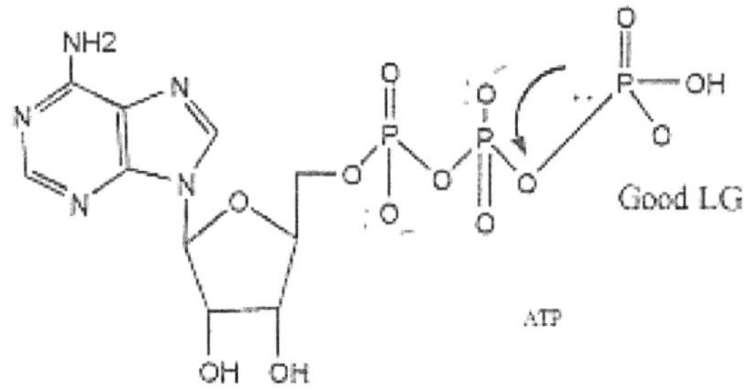

Good LG

ATP

ATP has a good leaving group which creates the electrophile.
Deprotonated glycerol becomes a nucleophile which attacks the Pi.

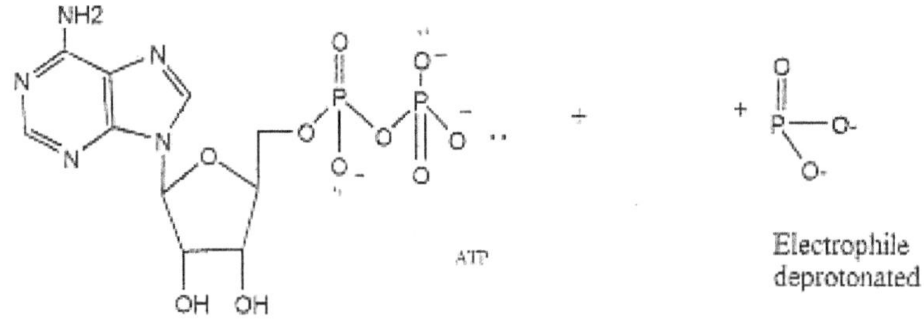

ATP

Electrophile
deprotonated

The kinase reaction is irreversible.

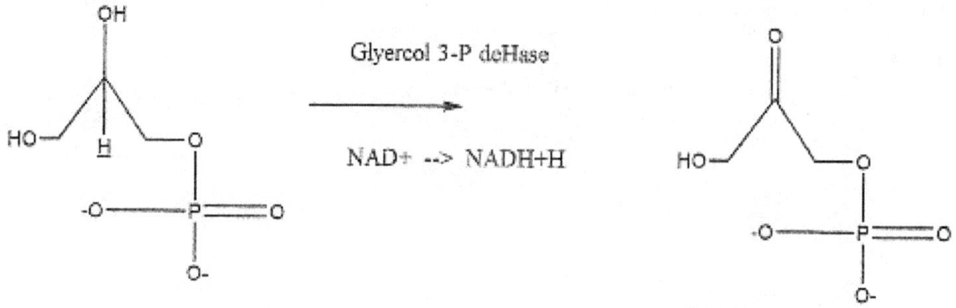

Glyercol 3-P deHase

NAD+ --> NADH+H

DHAP
Ketone

NAD+

NADH

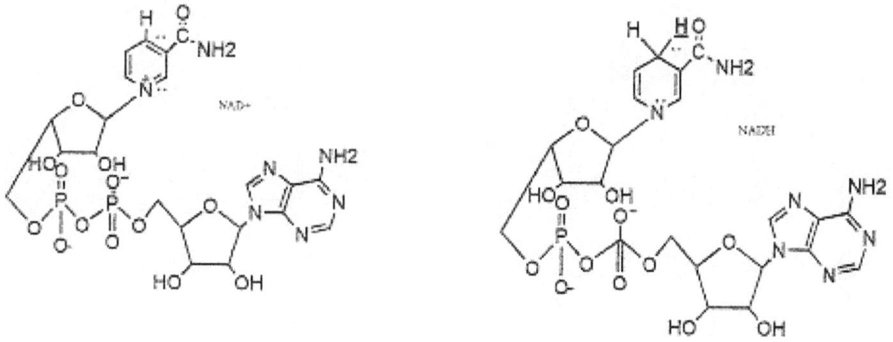

I refer you back to the G 3-PdeHase reaction for the details.

I invite you to write the details of the reactions for this intermediate.

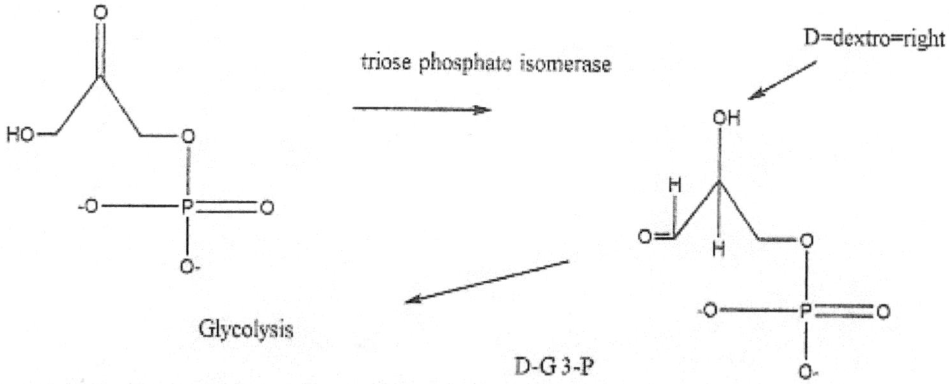

triose phosphate isomerase

D=dextro=right

Glycolysis

D-G 3-P

Again I refer and invite you to review the glycolysis triose phospate isomerase reaction.

In order to be oxidized, the f.a. must get to the mitochondrial matrix.
It crosses the inner and outer mitochondrial membranes.
It travels from the cytosol --> intermembrane space --> matrix.
The mechanism is called the carnitine shuttle.
There are 3 steps:
1. Esterification to CoA. **Carnitine acyltransferase I**
2. Transesterification to carnitine followed by transport. **acyl-carnitine/carnitine transporter**
3. Transesterification back to CoA. **carnitine acyltransferase II**

The chemical compounds in the reactions:
1. ATP
2. F.a.
3. Coenzyme A
3. Fatty acyl-adenylate (enzyme-bound)
4. Fatty acyl-CoA
5. Carnitine

ATP

Fatty acid

R=Long Chain Alkyl group
very stable
not involed in the reaction

Coenzyme A with thiol group

Carnitine

In mitochondrial cytosol

ATP + f.a. --> fatty acyl-adenylate (enzyme- bound)

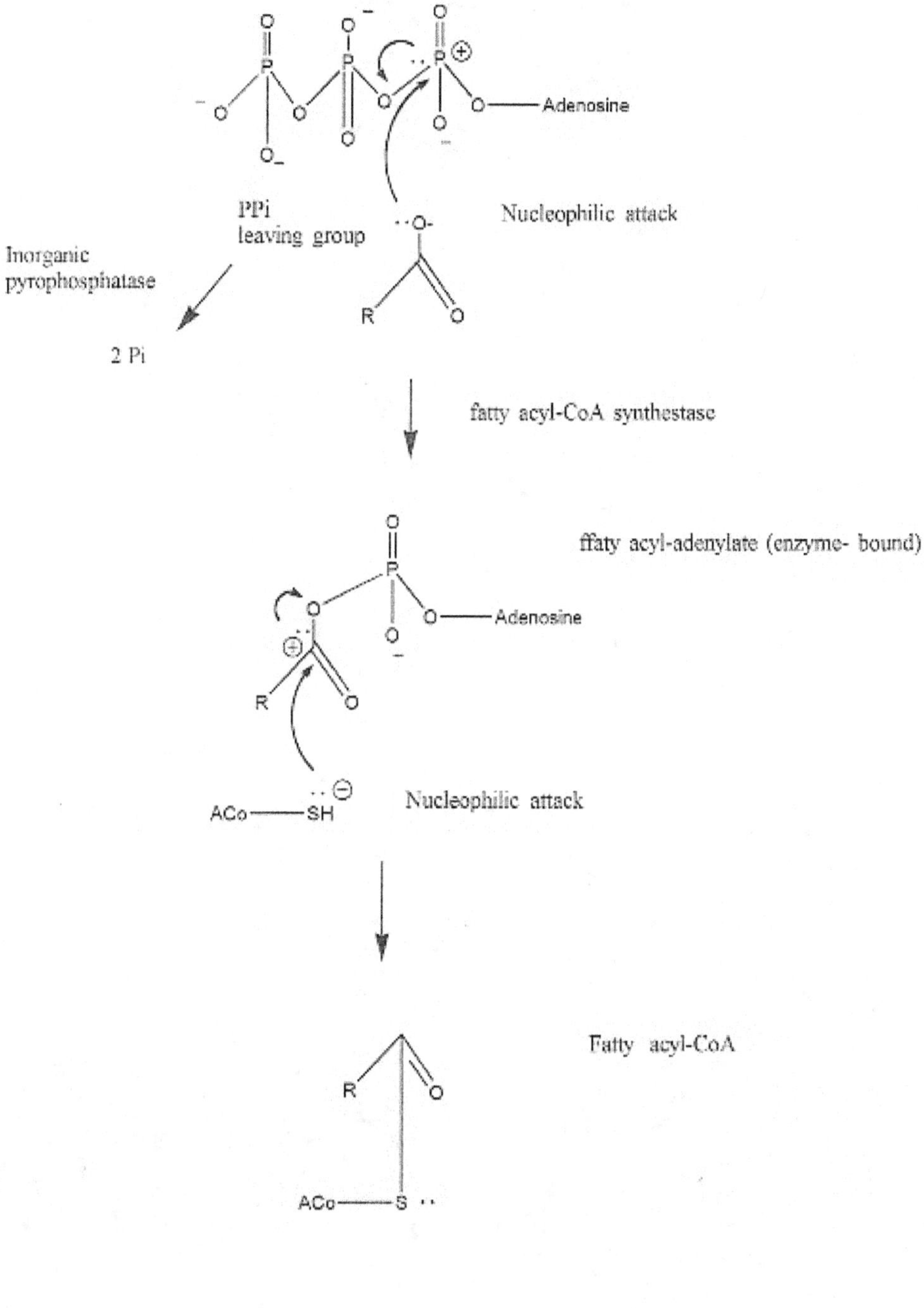
Adenosine
PPi
leaving group
Nucleophilic attack
Inorganic
pyrophosphatase
2 Pi
fatty acyl-CoA synthestase
ffaty acyl-adenylate (enzyme- bound)
Adenosine
ACo
SH
Nucleophilic attack
Fatty acyl-CoA
ACo

At the Outer Mitochondrial Membrane

Carnitine acyltransferase I is an integral membrane protein with enzyme activity.

Porins, membrane proteins, permit diffusion of carnitine.

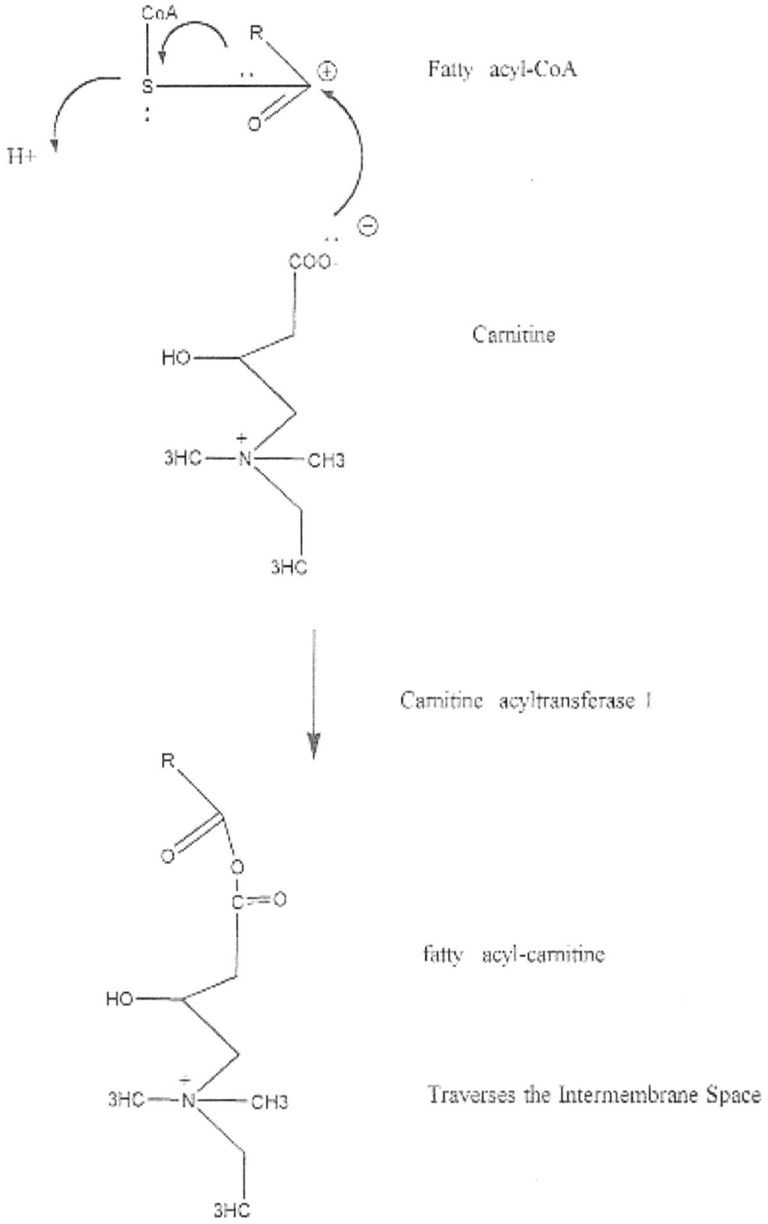

Fatty acyl-CoA

Carnitine

Carnitine acyltransferase I

fatty acyl-carnitine

Traverses the Intermembrane Space

At the Inner Mitochondrial membrane

Acyl-carnitine/carnitinettransporter delivers fatty acyl-carnitine to the Matrix

Carnitine acyltransferase 2 (integral protein) fatty acyl to CoA again.

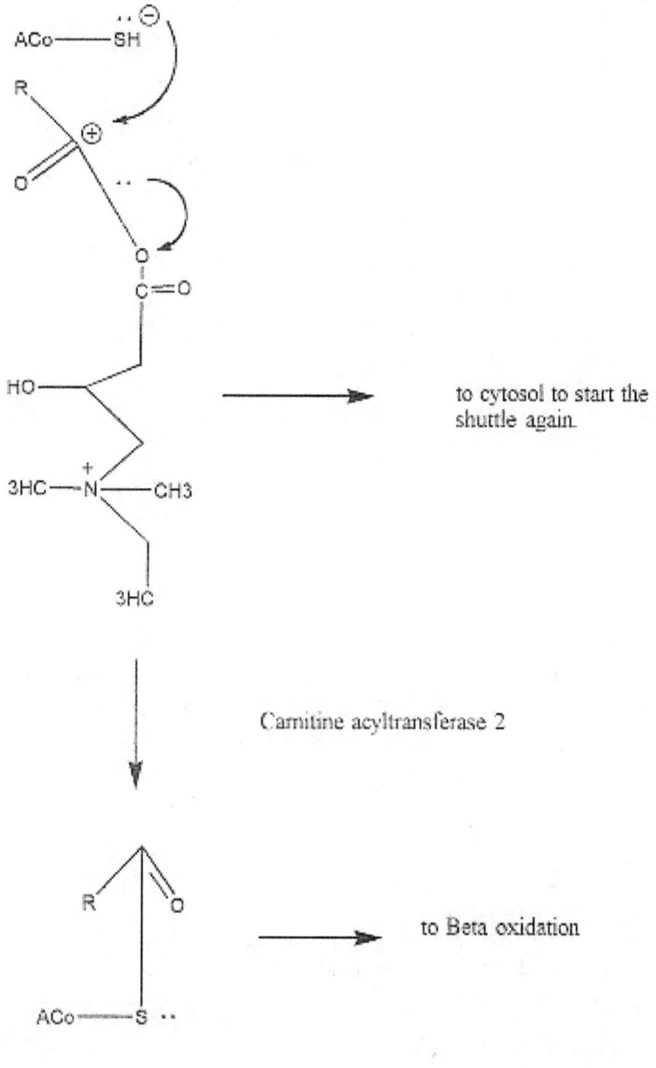
ACo—SH
R
O
O
C=O
HO
3HC—N—CH3
3HC
to cytosol to start the shuttle again.
Carnitine acyltransferase 2
R
O
ACo—S
to Beta oxidation

With the fatty acid energized by the CoA, it is ready for oxidation (releasing all the potential energy stored in the fatty acid). It is in the proper location for Beta-oxidation.

The 4 steps to Beta Oxidation of Saturated Fatty Acids

Beta = #3 C atom from carbonyl #1 C atom (-C=O).
Oxidation=lost of electrons
Saturation=all C atoms have single bonds and are hydrogenated (single bonds with H atoms).

We will assume that our f.a. is saturated.

General Reaction Highlights

C16 f.a.-CoA -->C14 f.a.-CoA + Acetyl-CoA with repetition.
Acetyl-CoA --> Citric Acid Cycle

by product: FADH2 --> ETF (electron-transferring Flavoprotein): mitochondrial repiratory chain, where the electrons are transferred to oxygen.

by product: NADH + H+ --> NADH dehydrogenase --> respiratory chain

Reactants: H2O + CoA-SH

The last 3 steps are done by TFP (trifunctional Protein) enzyme for C chains 12 or greater.

TFP is a hetero - octomer: alpha 4 Beta 4: alpha: hydratase (H2O addition) and dehydrogenase (removal of H atoms).   beta: thiolase (Addition of CoA-SH).
This enzyme facilitates efficient substrate channeling (tight substrate control around active sites).

So Beta oxidation extracts electrons to produce ATP (energy) for physiological reactions.
For the football player, these could be nervous, cardiovascular, muscle, pulmonary, renal. and skin function. Nervous: Alertness and Mentation
        Cardiovascular: BP and O2 transport to vital organs
        Muscle: Excercise
        Pulmonary: Breathing requires muscle activity.
        Renal: Handling of Na+ loads.
        Skin: Sweating

We will describe each step demonstrating the organic chemistry principles.

Reaction 1

Acyl-CoA + FAD - acyl-CoA dehydrogenase --> trans-delta^2-Enoyl-CoA + FADH2
 e source   e sink
                prosthetic group

Reduced acyl-CoA dehydrogenase, via FADH2, donates electrons to ETF (electron carrier of respiratory chain).   e source   e sink

Electrons from respiratory chain pass to O2.
                e source   e sink-final

1.5 ATP molecules per electron pair: 2 e = 1.5 ATP molecules

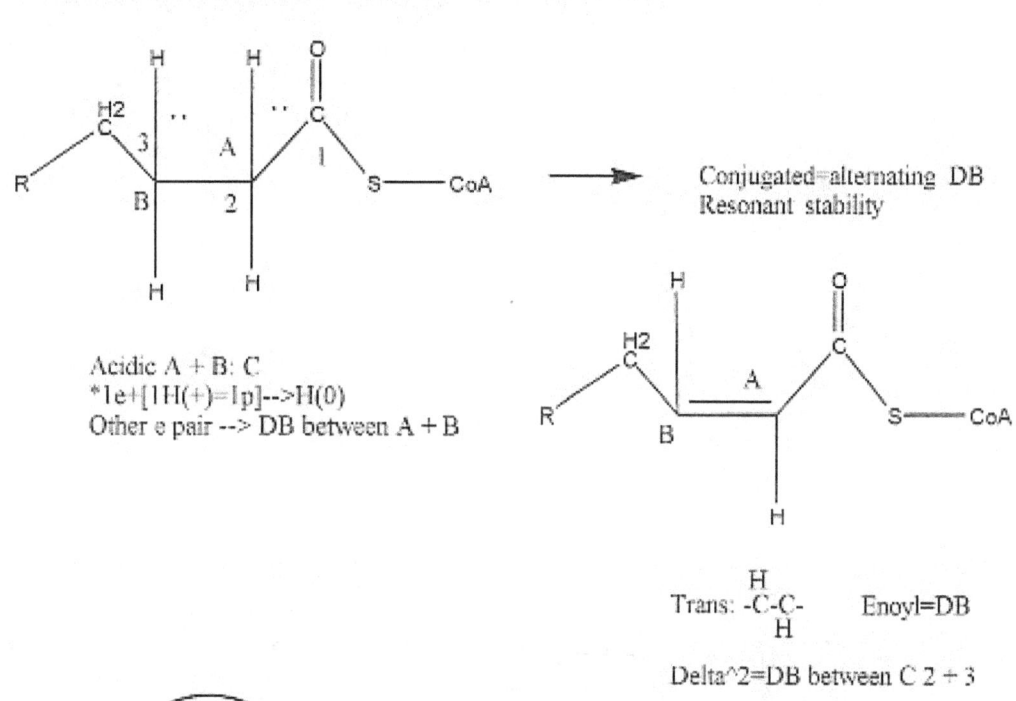

Acidic A + B: C
*1e+[1H(+)=1p]-->H(0)
Other e pair --> DB between A + B

Conjugated=alternating DB
Resonant stability

Trans: -C-C-   Enoyl=DB

Delta^2=DB between C 2 + 3

FAD ⟶ FADH2

Clarification

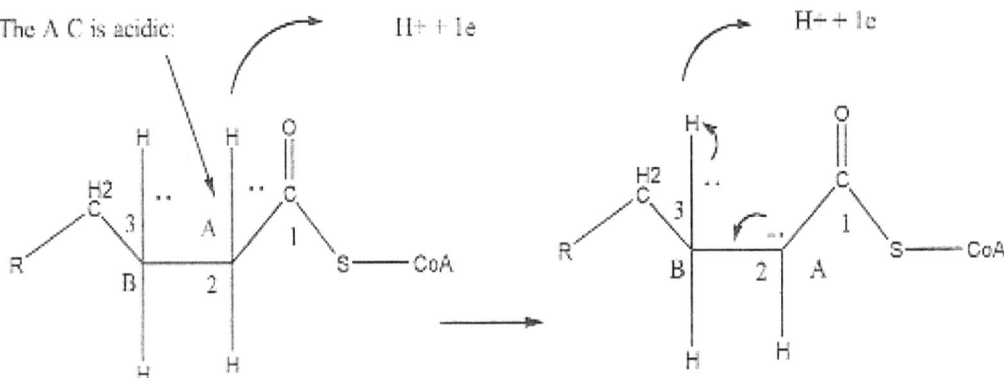

The A C is acidic:

H+ + 1e

H− + 1e

The B C is acidic by induction

Conjugation: Alternating DB

Resonance 2e can move through neighboring atomic orbitals --> stability

FAD

FMN

FAD

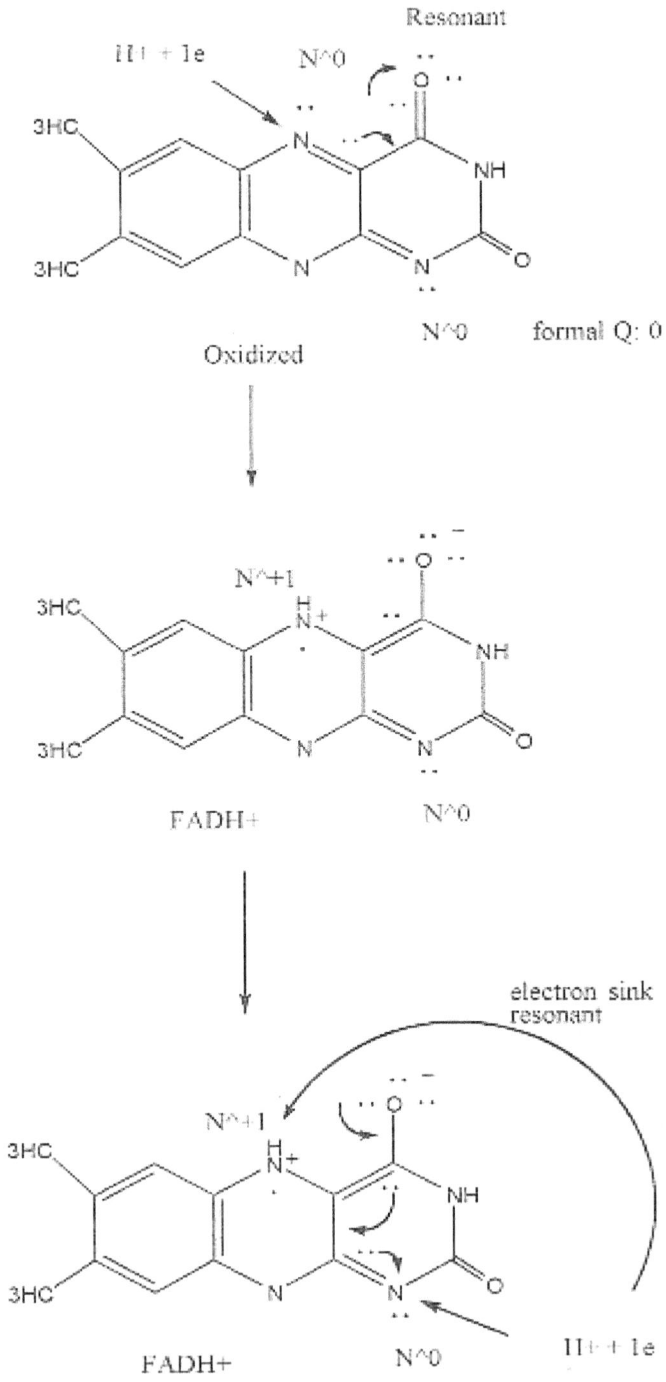

Resonant

H+ + 1e

N^0

3HC

3HC

NH

O

N^0

formal Q: 0

Oxidized

N^+1

3HC

H+

3HC

NH

O

N

N^0

FADH+

N^0

electron sink
resonant

N^+1

3HC

H+

3HC

NH

O

semiquinone

N

N^0

H+ + 1e

FADH+

N^0

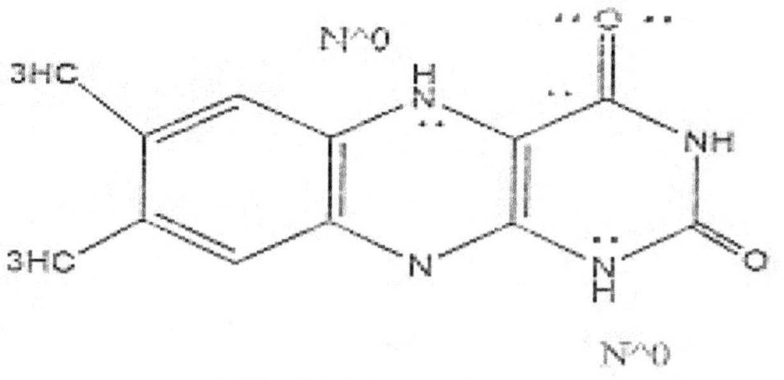

3HC

3HC

N^O

N^O

FADH2
reduced

Reaction 2

Addition of H2O across the DB

enoyl-CoA hydratase

⊖ : The A C is partially (-) because of induction by Oxygen.

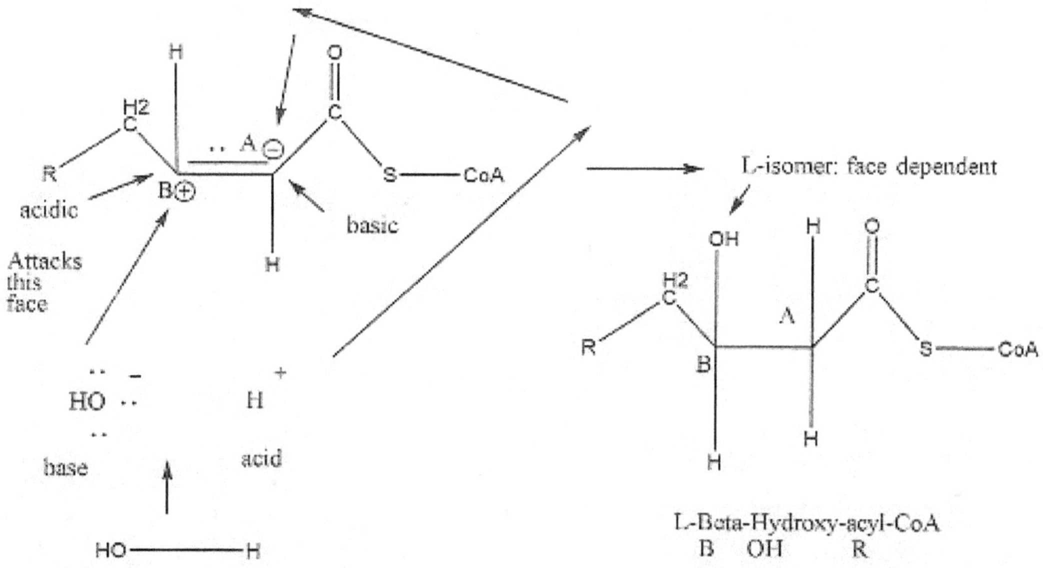

R

acidic

Attacks
this
face

HO ··

base

HO————H

H2
C

B⊕

basic

H

A⊖

C

S———CoA

H

acid

L-isomer: face dependent

OH

H2
C

R

B

H

A

H

C

S———CoA

L-Beta-Hydroxy-acyl-CoA
B   OH   R

Enoyl can be involved in acid base reaction
Base accepts H+
Acid accepts OH-

Removal of H2 from Beta Carbon: H+ + H:-

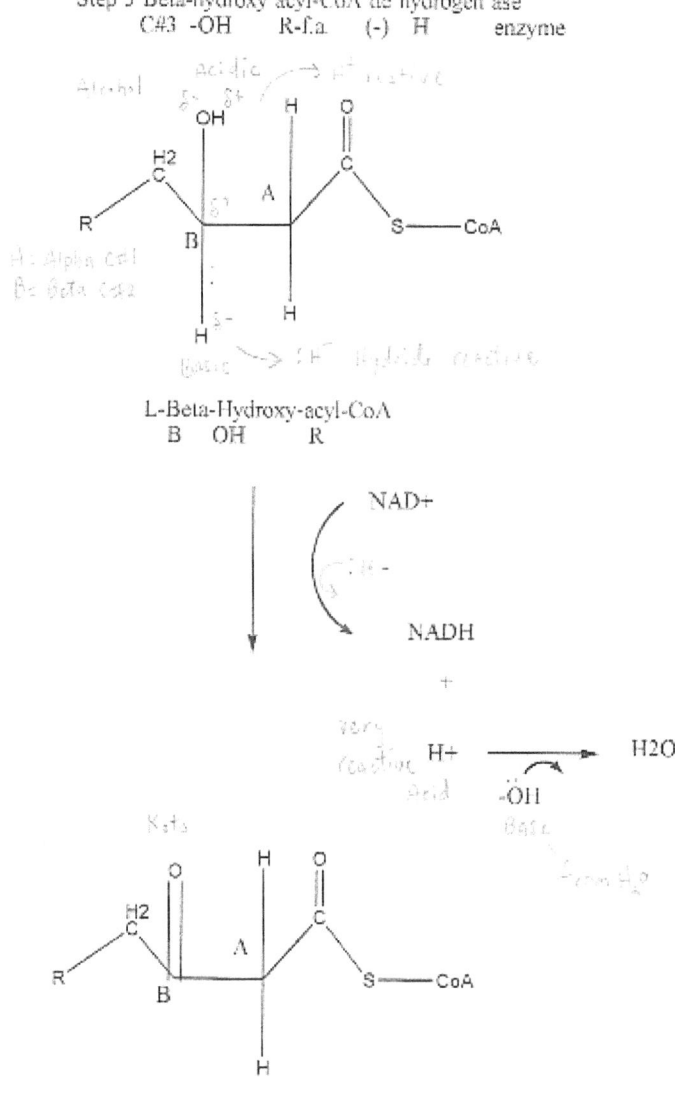

Step 3 Beta-hydroxy acyl-CoA de hydrogen ase

C#3 -OH   R-f.a   (-) H   enzyme

L-Beta-Hydroxy-acyl-CoA
B   OH   R

NAD+

NADH

+

H+ ———→ H2O

Beta-Keto acyl-CoA
C#3 C O R

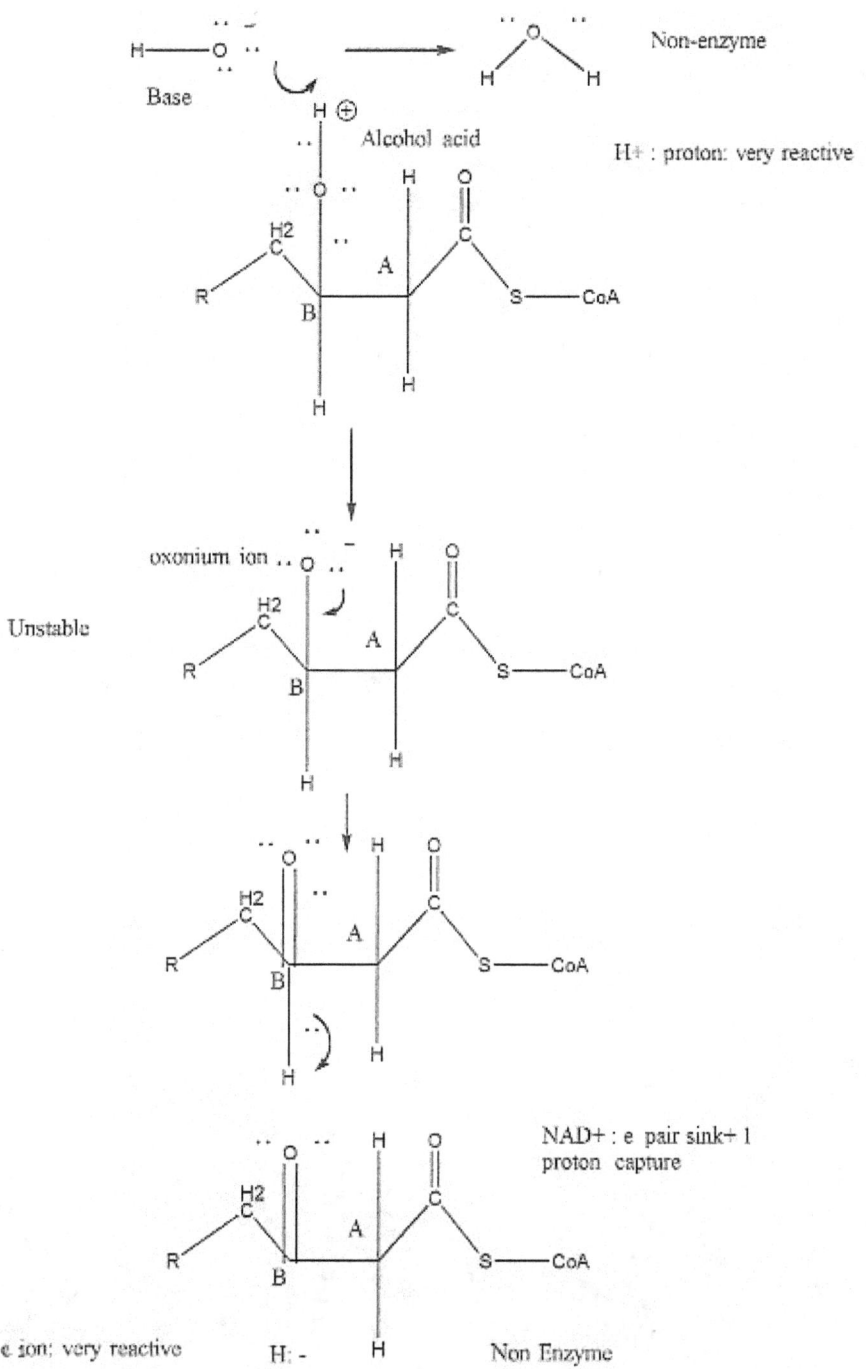

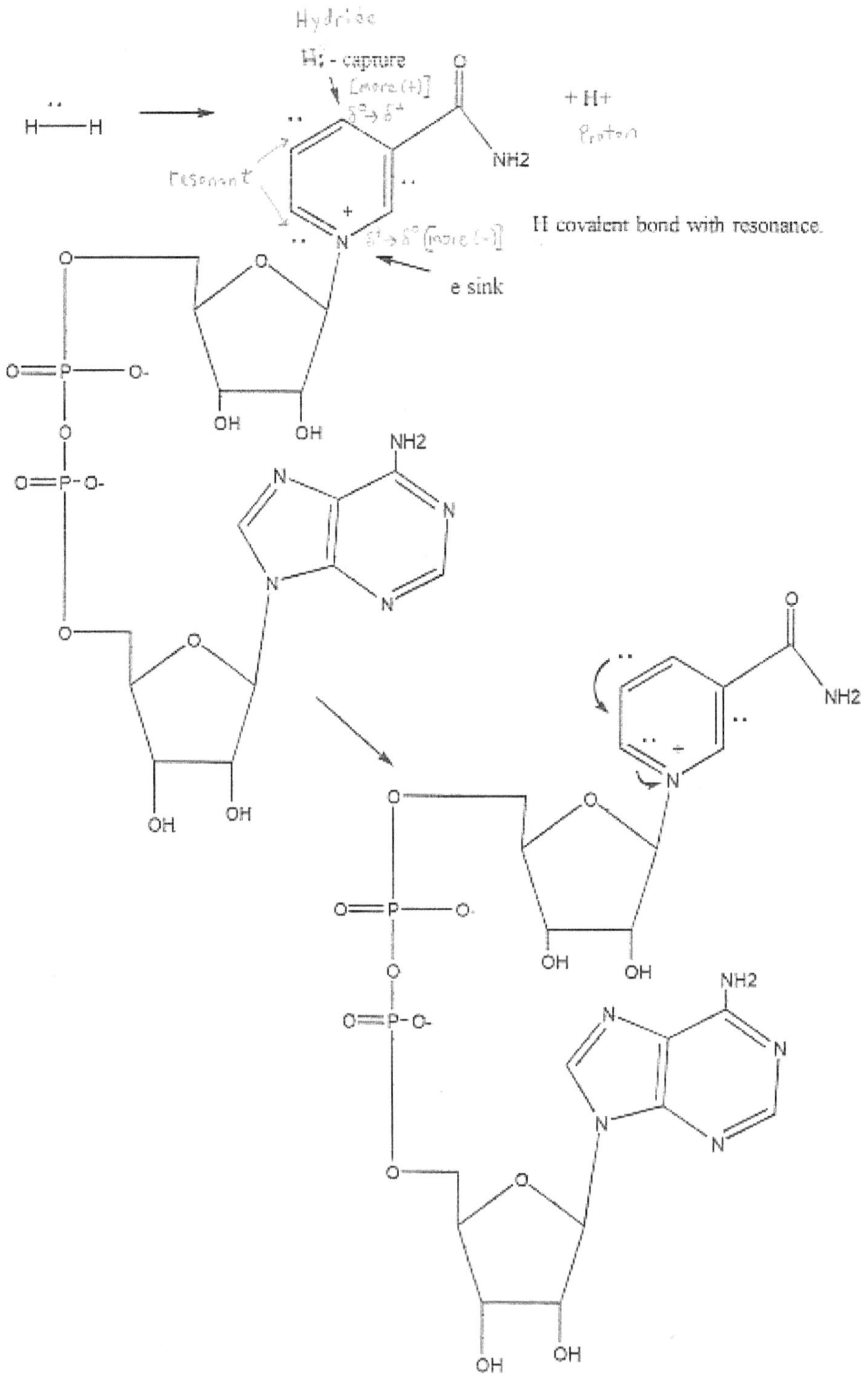

Hydride

H: - capture
[more.(+)]
$\delta^- \to \delta^+$

+ H+
Proton

Resonant

H covalent bond with resonance.

e sink

Hydride Addition H:⁻

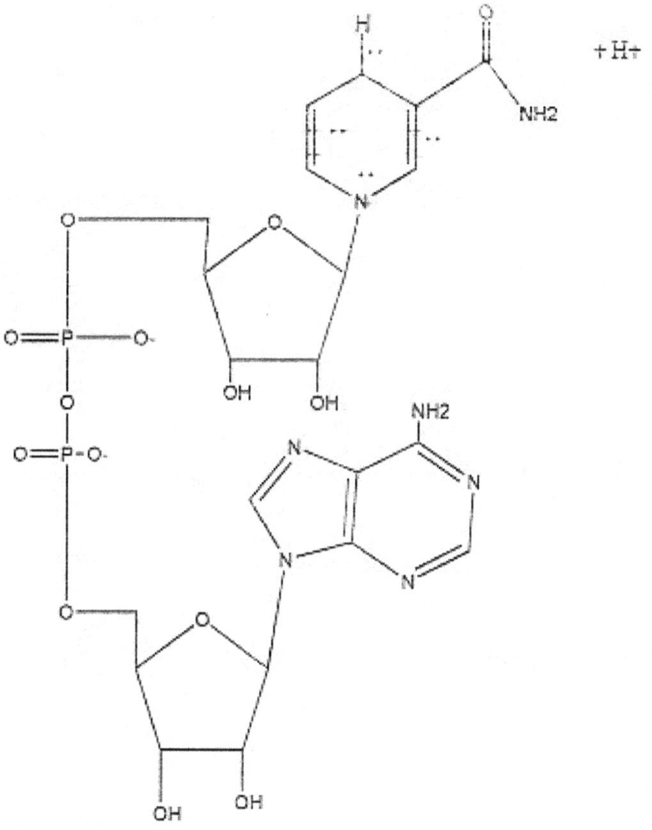

+H+

Step 4: Split of Beta-alpha bond by CoASH (Nucleophilic attack)

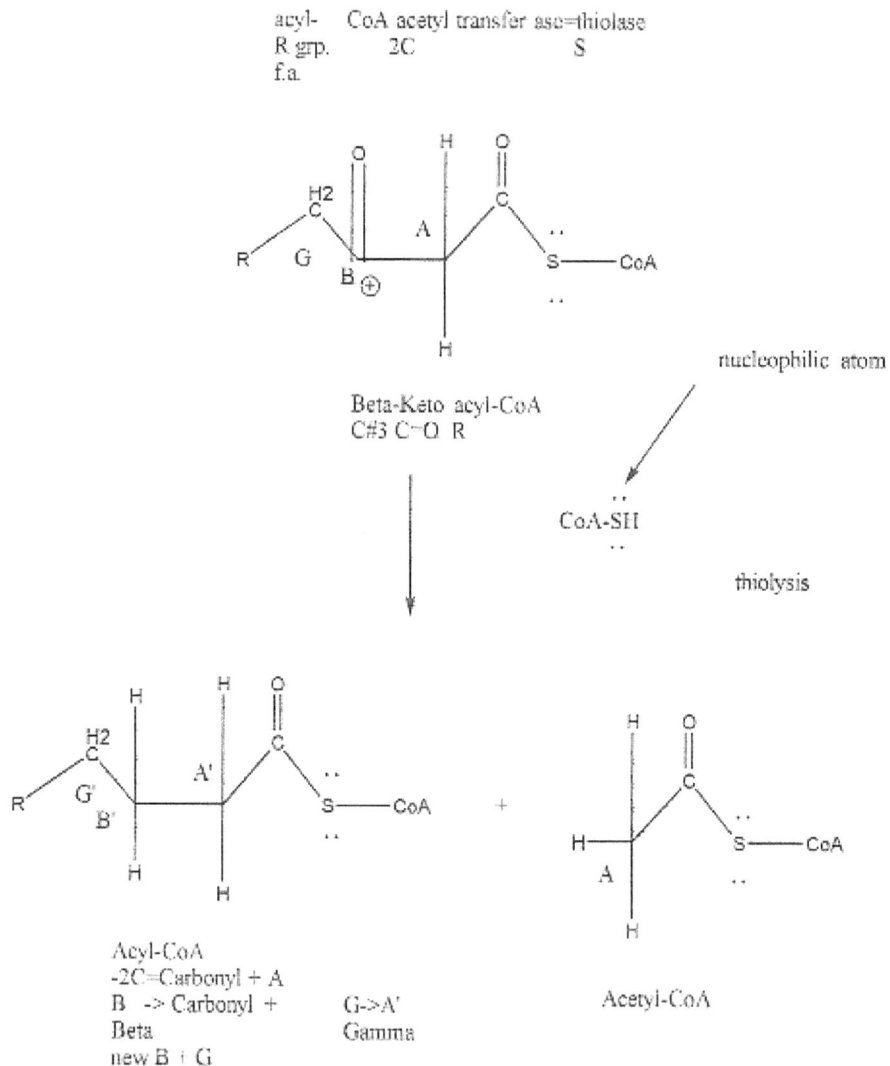

Beta-Keto acyl-CoA
C#3 C=O R

nucleophilic atom

CoA-SH

thiolysis

Acyl-CoA
-2C=Carbonyl + A
B -> Carbonyl +   G->A'
Beta   Gamma
new B + G

Acetyl-CoA

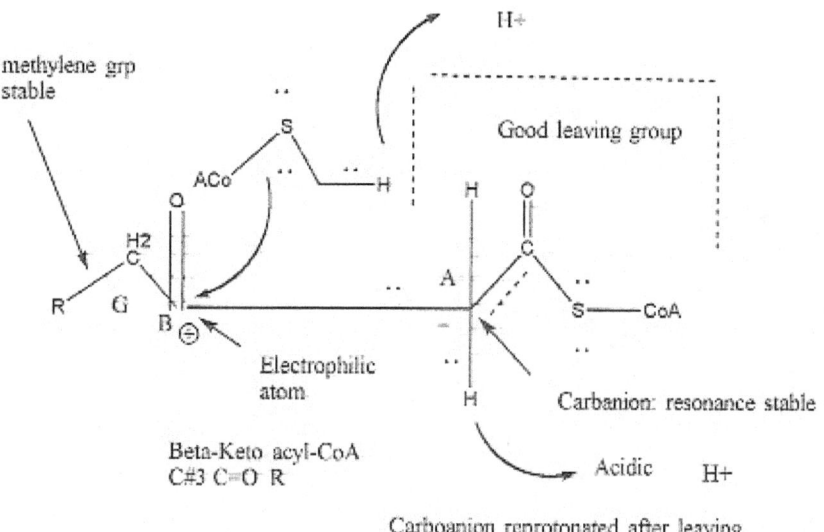

methylene grp stable

Good leaving group

H+

Electrophilic atom

Carbanion: resonance stable

Beta-Keto acyl-CoA
C#3 C=O R

Acidic   H+

Carboanion reprotonated after leaving.

B   A
—C——C—   Methylene group: stable
 H2   H2

Polarity=electrostatic attraction/repulsion

Unstable

B   A

For completeness: F.a. C# >/= 12 -> TFP (trifunctional protein=multienzyme complex in the inner mitochondrial membrane), heterooctamer=A4B4, efficient substrate channeling, keep close to surface.

C# </= 12 -> matrix.

Clarification of Details:

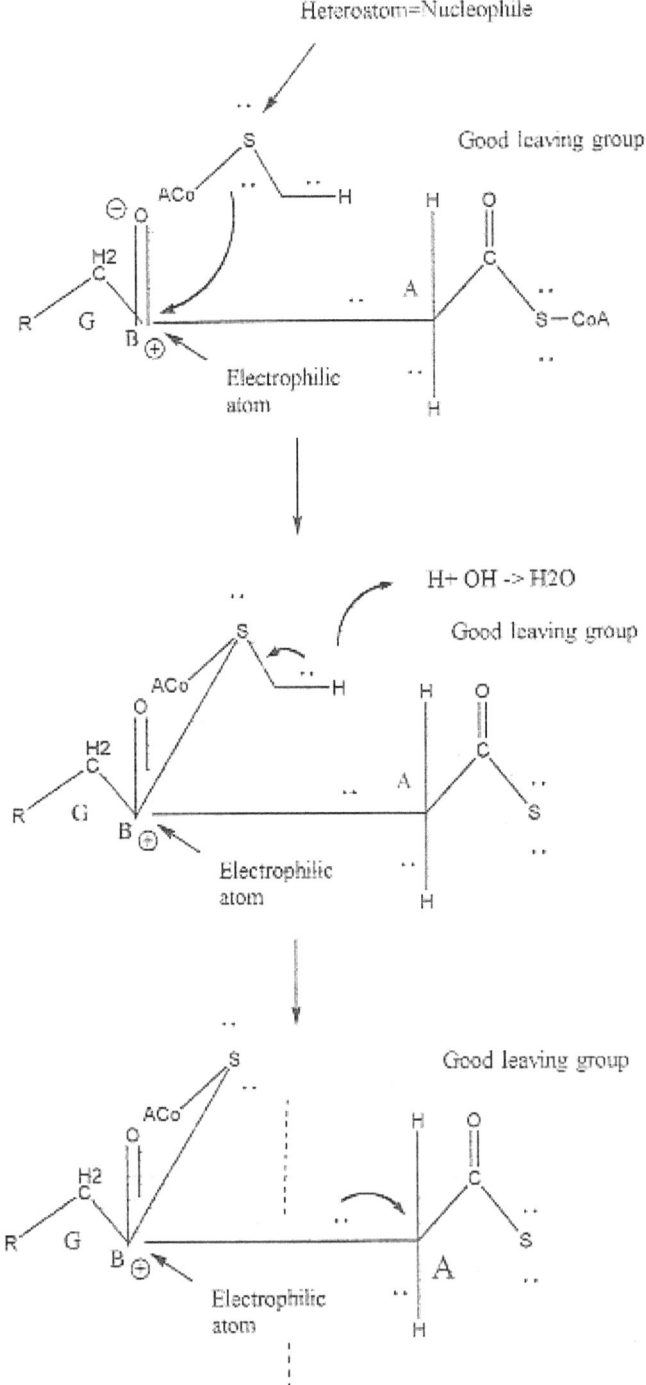

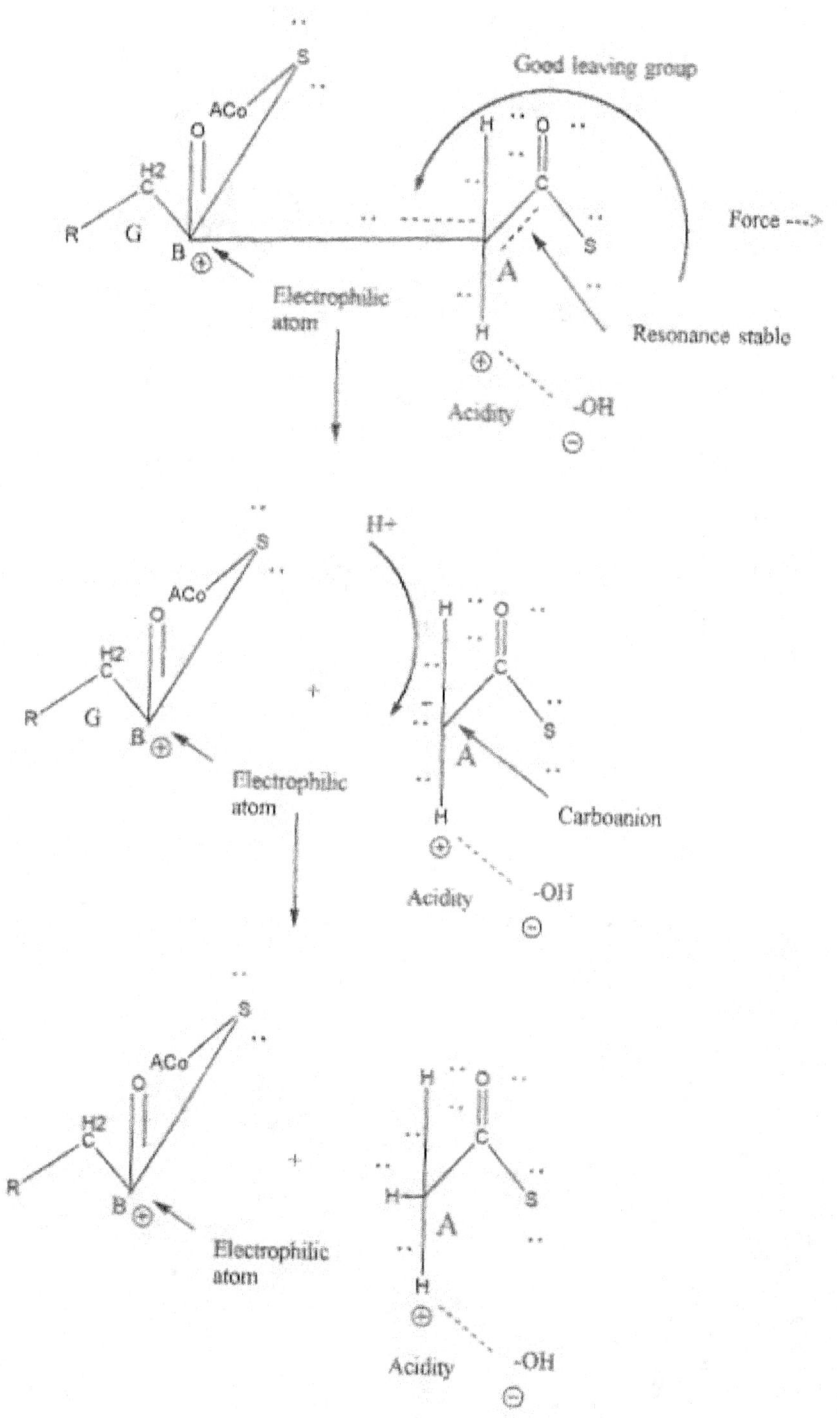

Coenzyme A = CoASH

Chapter 9

Energy 3

This chapter will cover our 3rd nutrient-protein. As mentioned, protein is digested and absorbed as amino acids-the building blocks or monomers of protein. To com bust or burn amino acid is energy costly. It requires energy to transform the amino acid to glucose. This process is called Gluco- neo-genesis. The creation of glucose from non-carbohydrate sources. The order of preference in the creation of energy is as follows:

Carbohydrate > Lipids > Protein.

In periods of stress, such as starvation or exercise, amino acid become a source of energy. Why? The brain during normal states utilize ½ of the stored glucose in liver and muscle, about 120 grams/day.

The place where this genesis takes place is mainly the liver and renal cortex.

3 or 4 Carbon structured sources are used: Lactate, pyruvate, glycerol, and amino acid. For example, in the **Cori** cycle, lactate is built up in the muscle during exercise. Then this is transported to the liver and converted to glucose. From there, it is sent to the muscle and stored as glycogen.

The overall scheme of amino acid conversion to glucose is:

Glucogenic a.a. $\rightarrow$ TCA $\rightarrow$ PEP $\rightarrow$ G6P

Many of the steps are from the reversible reactions of glycolysis. They have a ΔG near 0. There are 3 irreversible steps which require new reactions to reverse the process (bypass). Why they are irreversible is because energy in the form of ATP is spent to drive the reaction forward, creating a negative free energy change. These are committed steps and regulated ones of the metabolic pathway. This promotes efficient channeling of the intermediates to produce the proper end-product. The expenditure of ATP overcomes the energy barrier of the forward reaction; thus ensuring the reaction will take place. $-\Delta G$ is a math construct having the same meaning. It is a thermodynamic entity which we will learn later. The energy lies in the 3 PO4 -2 bonds of ATP. When they break, they release large amounts of energy from the reaction. These reactions are exergonic. There is an increase in potential energy in the new compound by PO4 -2 group transfer. This takes place in the cytosol.

The 3 irreversible steps are:

1. Glucose $\rightarrow$ G6P (Hexokinase)   -1 ATP
2. F6P $\rightarrow$ F 1,6 BP (Phosphofructokinase-1)   -1 ATP
3. 2PEP $\rightarrow$ 2Pyruvate (Pyruvate kinase)   +2 ATP

Let us trace the steps of a typical amino acid, alanine, during gluconeogenesis.

In the mitochondria: (Liver)

Alanine → pyruvate → oxaloacetate (pyruvate carboxylase)   -2ATP exergonic -ΔG

Transamination   coenzyme: biotin

Pyruvate + HCO3 - + ATP → oxaloacetate + ADP + Pi
        biotin

The transamination process: B6:

B6 combines with the enzyme in the form of a Schiff Base.
The process is a dehydration type; water is removed. This is detailed. By hydration, the reverse process, B6 can participate in the same reaction except this time with Ala.
Through a series of organic steps, Ala is transformed into pyruvate. Pyruvate can be converted to oxaloacetate, the first irreversible step in gluconeogenesis.

Notes:

Place: Mitochondria

Alkyl group

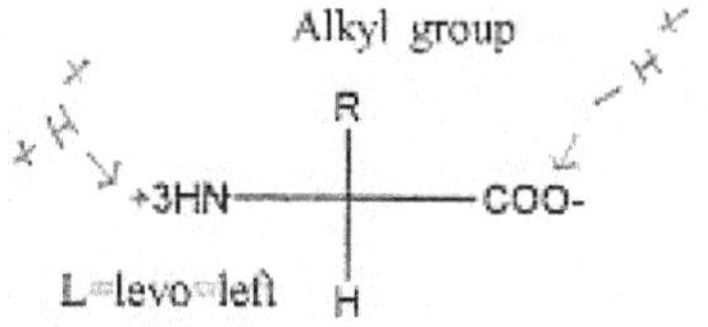

R

$+3HN$ —— COO-

L=levo=left   H

$+H^+$ $-H^+$

generic L-amino acid
ionized at physiologic pH   →   CH3

$+3HN$ —— COO-

H

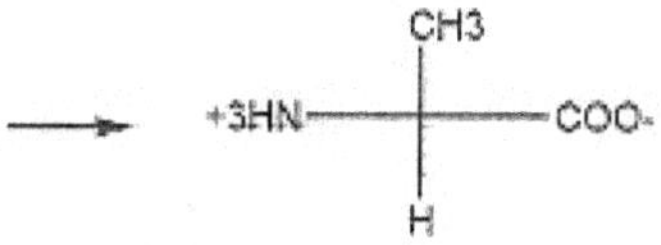

a.a.=amino carboxylic acid   Alanine=A=Ala

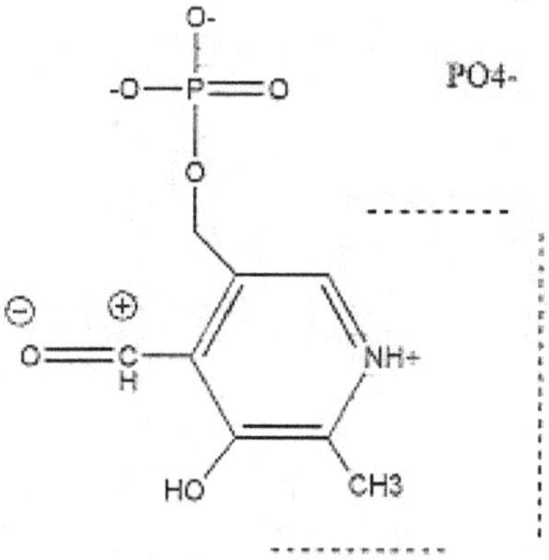

O-

-O—P=O

O

PO4-   PLP

Prostetic group=coenzyme
amino transfer ase
NH3+ group

$\ominus$ $\oplus$

O=C
H

Pyridoxine=B6

NH+

HO   CH3

O-

-O—P=O

O

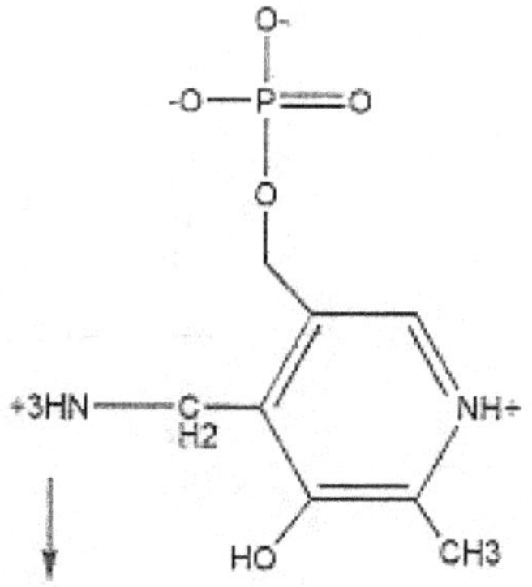

Pyridoxamine phosphate

$+3HN$—C
H2

NH+

HO   CH3

NH4+
ammonia

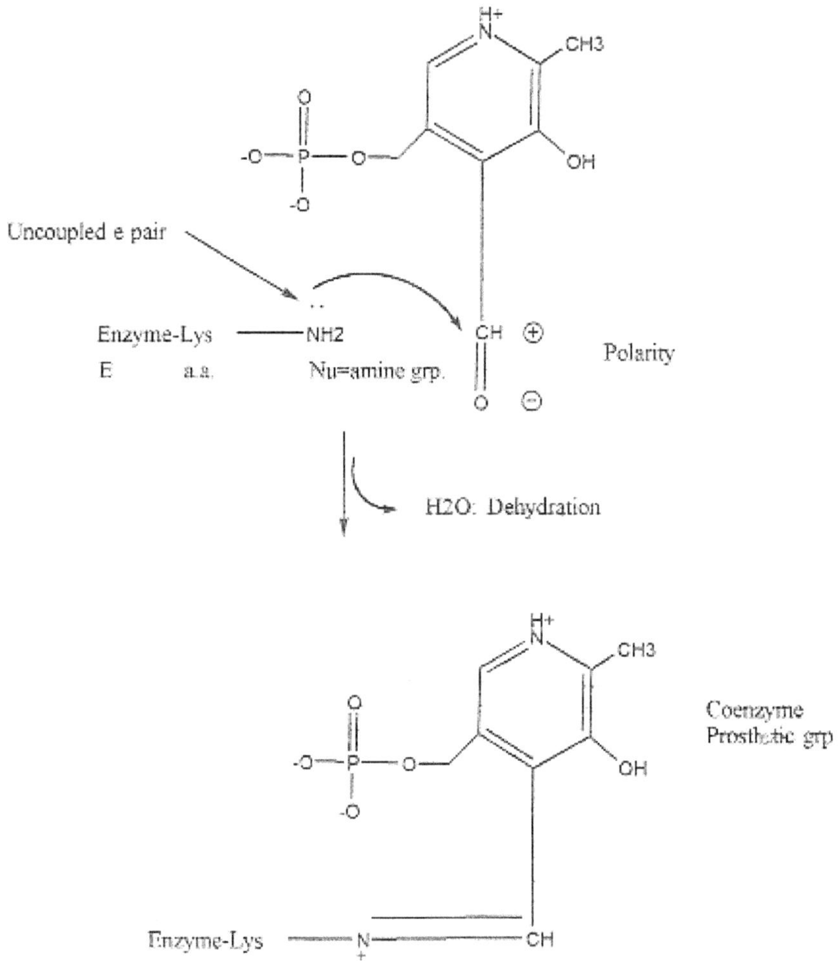
H+
N
CH3
O
-O—P—O
-O
OH
Uncoupled e pair
Enzyme-Lys
NH2
E
a.a.
Nu=amine grp.
CH
O
Polarity
H2O: Dehydration
Coenzyme
Prosthetic grp
Enzyme-Lys
N
CH

Schiff Base: Enzyme + B6

Amino transfer ase

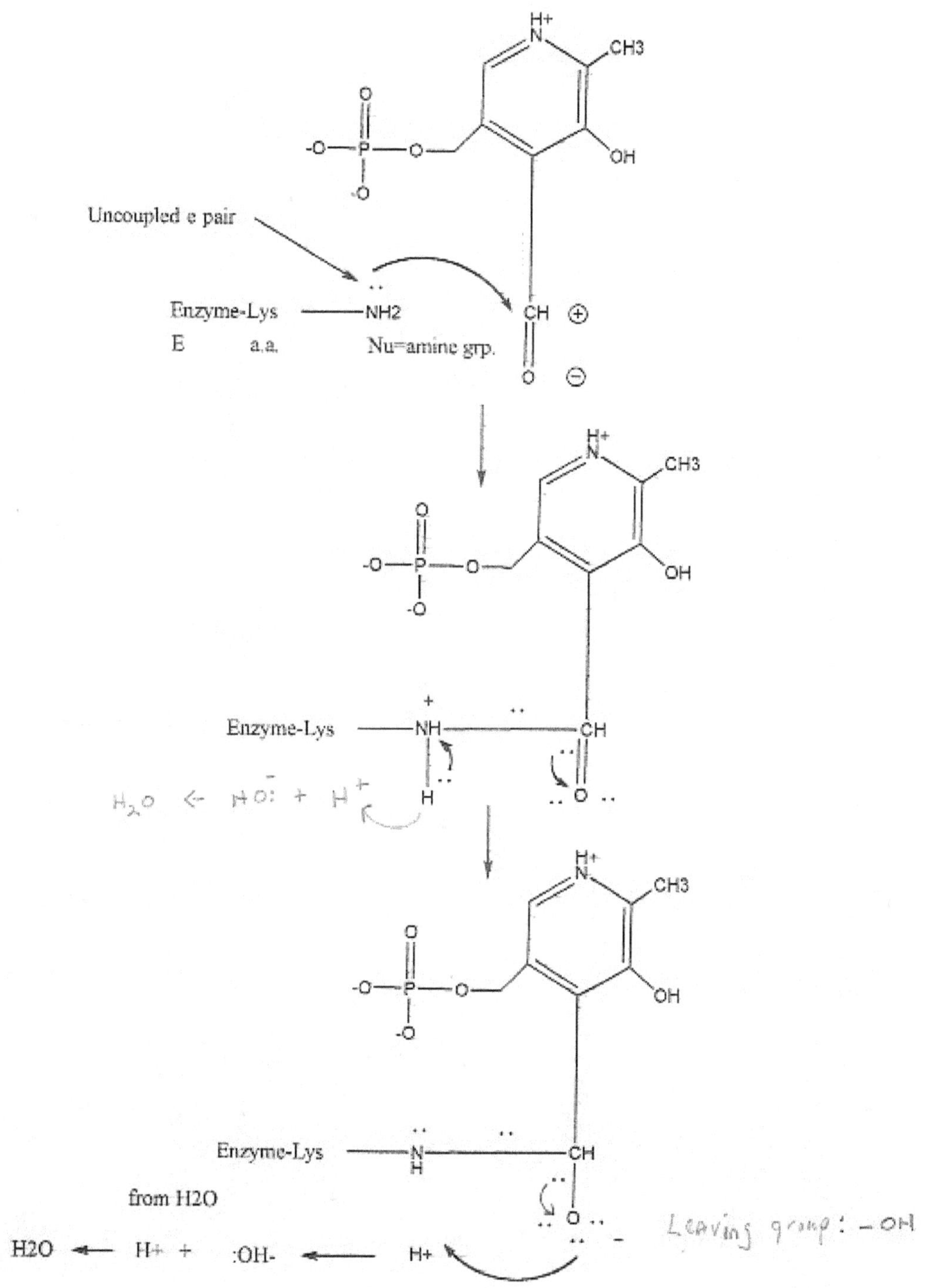
H+
CH3
OH
Uncoupled e pair
Enzyme-Lys
NH2
CH
E
a.a.
Nu=amine grp.
Enzyme-Lys
NH
CH
Enzyme-Lys
N
H
CH
from H2O
H2O
H+ +
:OH-
H+
Leaving group: -OH

Enzyme aldimine form to Alanine aldimine form.

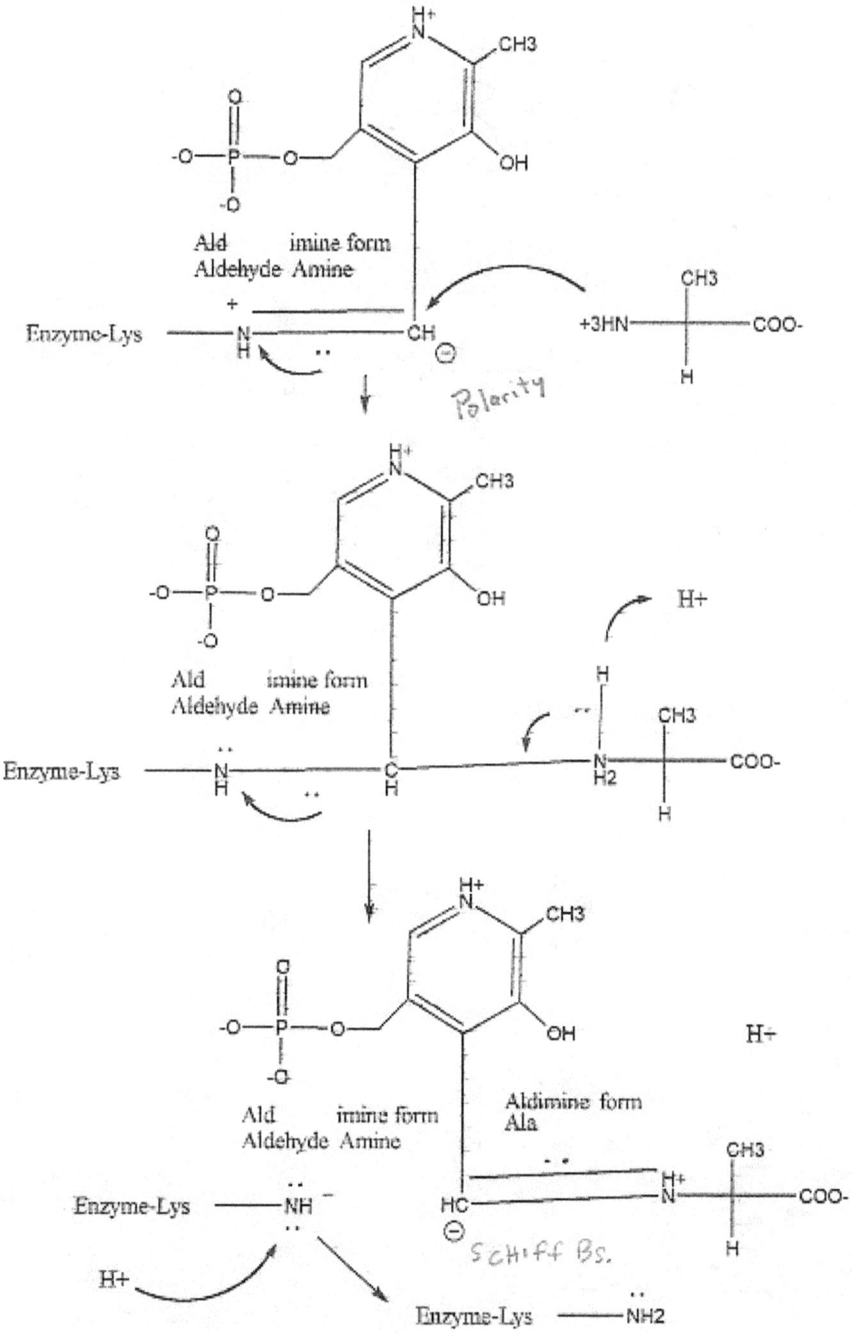

Ald   imine form
Aldehyde Amine

Enzyme-Lys

Polarity

Ald   imine form
Aldehyde Amine

H+

Enzyme-Lys

Ald   imine form
Aldehyde Amine

Aldimine form
Ala

H+

Enzyme-Lys ——NH⁻

H+

Schiff Bs.

Enzyme-Lys ——NH2

Resonance is the flow of electron potential throughout the molecule.

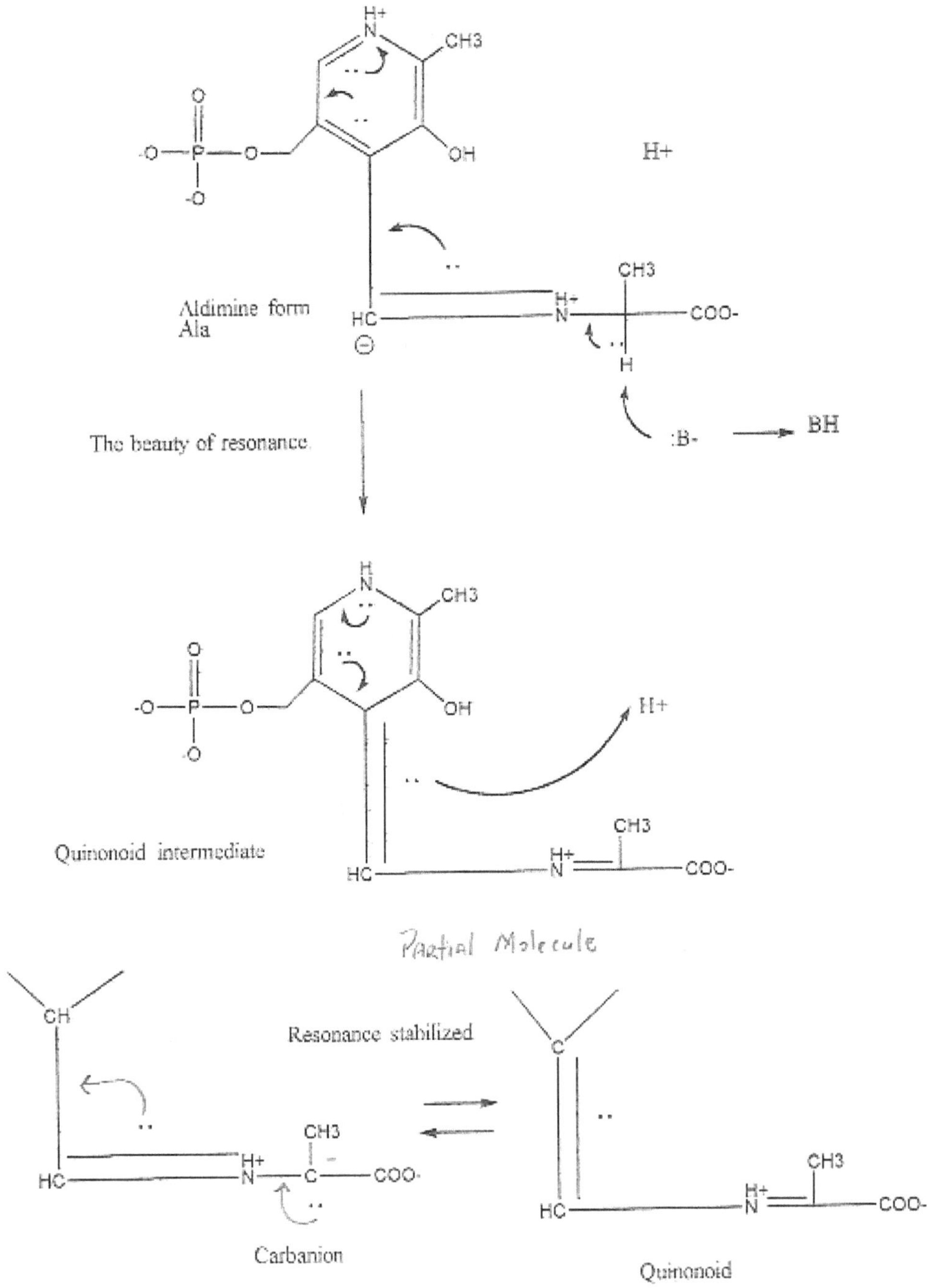

Aldimine form
Ala

H+

The beauty of resonance.

Quinonoid intermediate

Partial Molecule

Resonance stabilized

Carbanion

Quinonoid

Hydration is the addition of water.

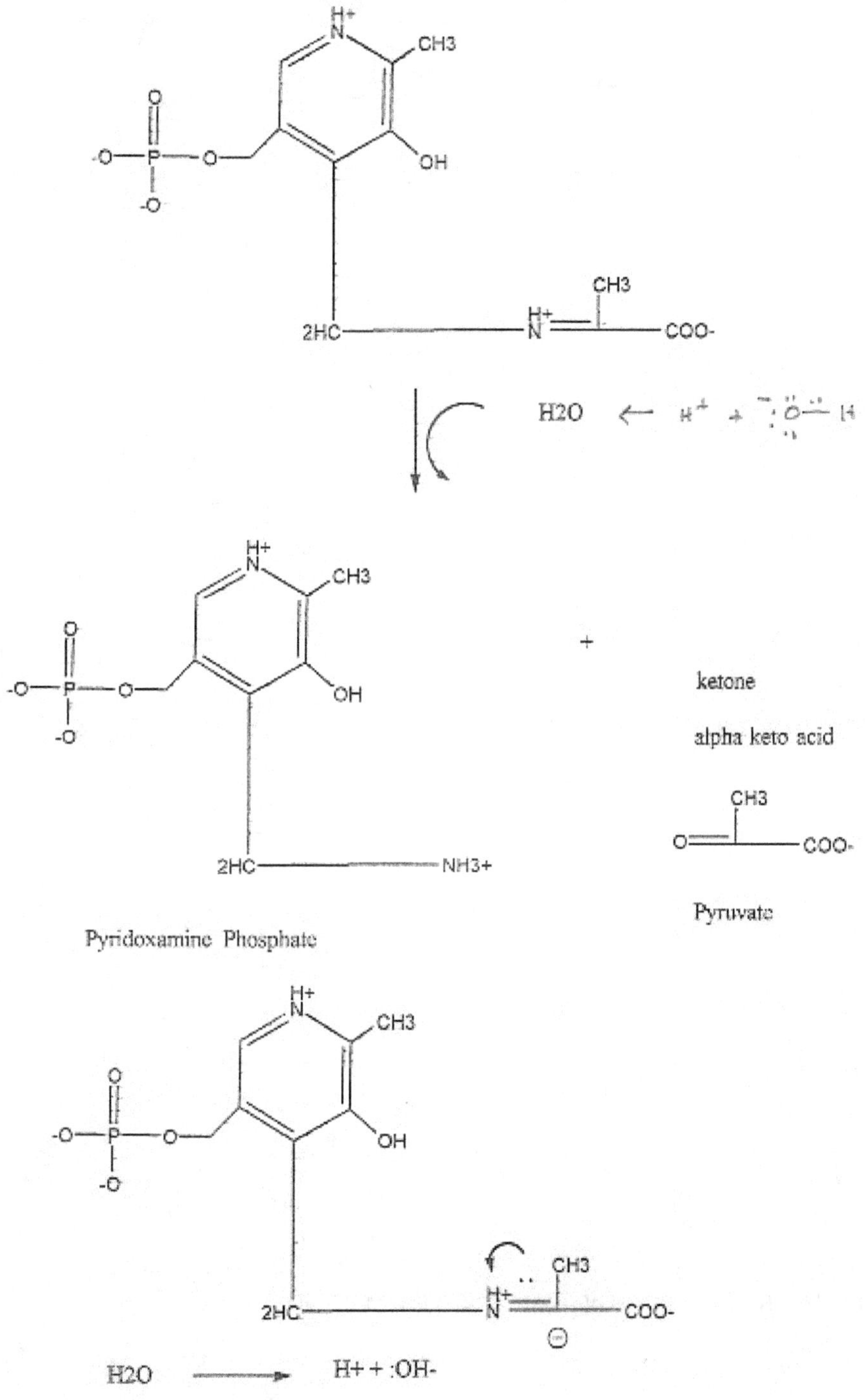

Pyridoxamine Phosphate

ketone

alpha keto acid

Pyruvate

$H2O \longrightarrow H+ + :OH-$

Alanine transfoms to pyruvate with the release of B6.

Ala + PLP -> Pyruvate+ aminated B6 (pyridoxamine)
L-a.a.   alpha-keto ac.

The other half of the transaminase reaction is:
pyridoxamine + alpha-keto glutarate -> L-Glutamine

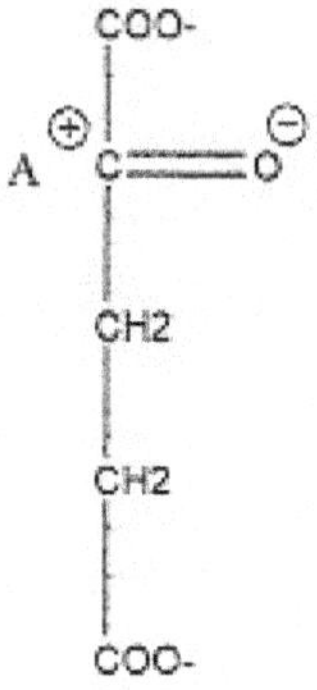
COO-
A
C
O
CH2
CH2
COO-

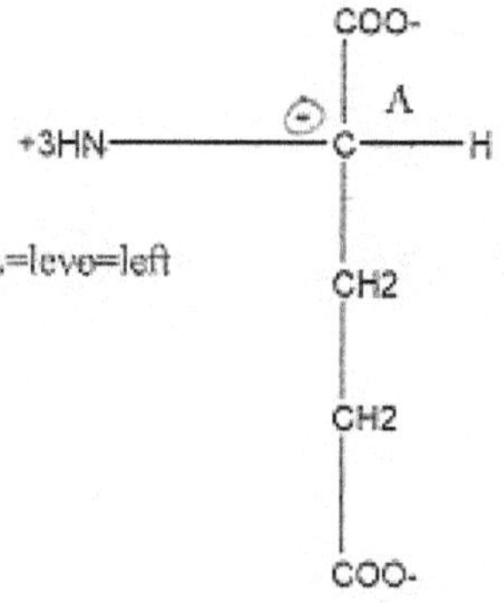
COO-
A
+3HN
C
H
L=levo=left
CH2
CH2
COO-

alpha-keto glutarate
A   ketone

L-Glutamate

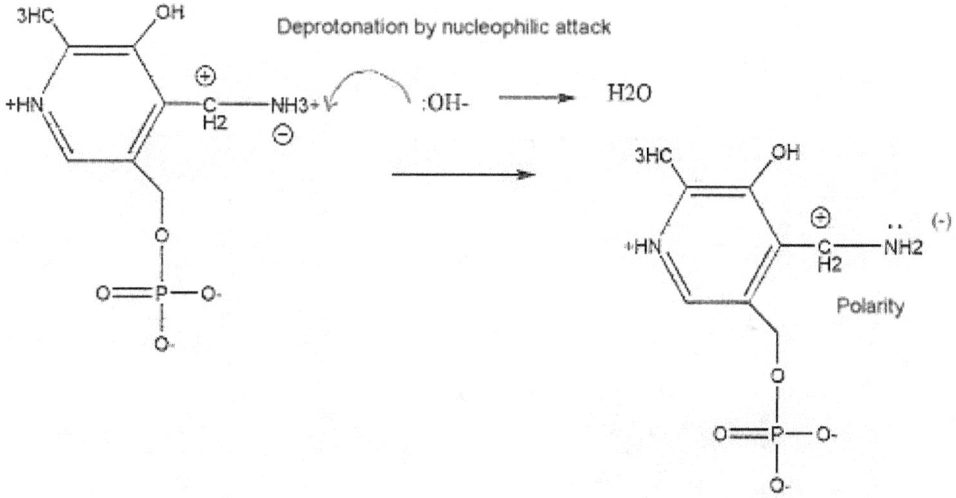
3HC
OH
+HN
C
H2
NH3+
Deprotonation by nucleophilic attack
:OH-
H2O
NH2
(-)
Polarity
O
O
P
O-
O-

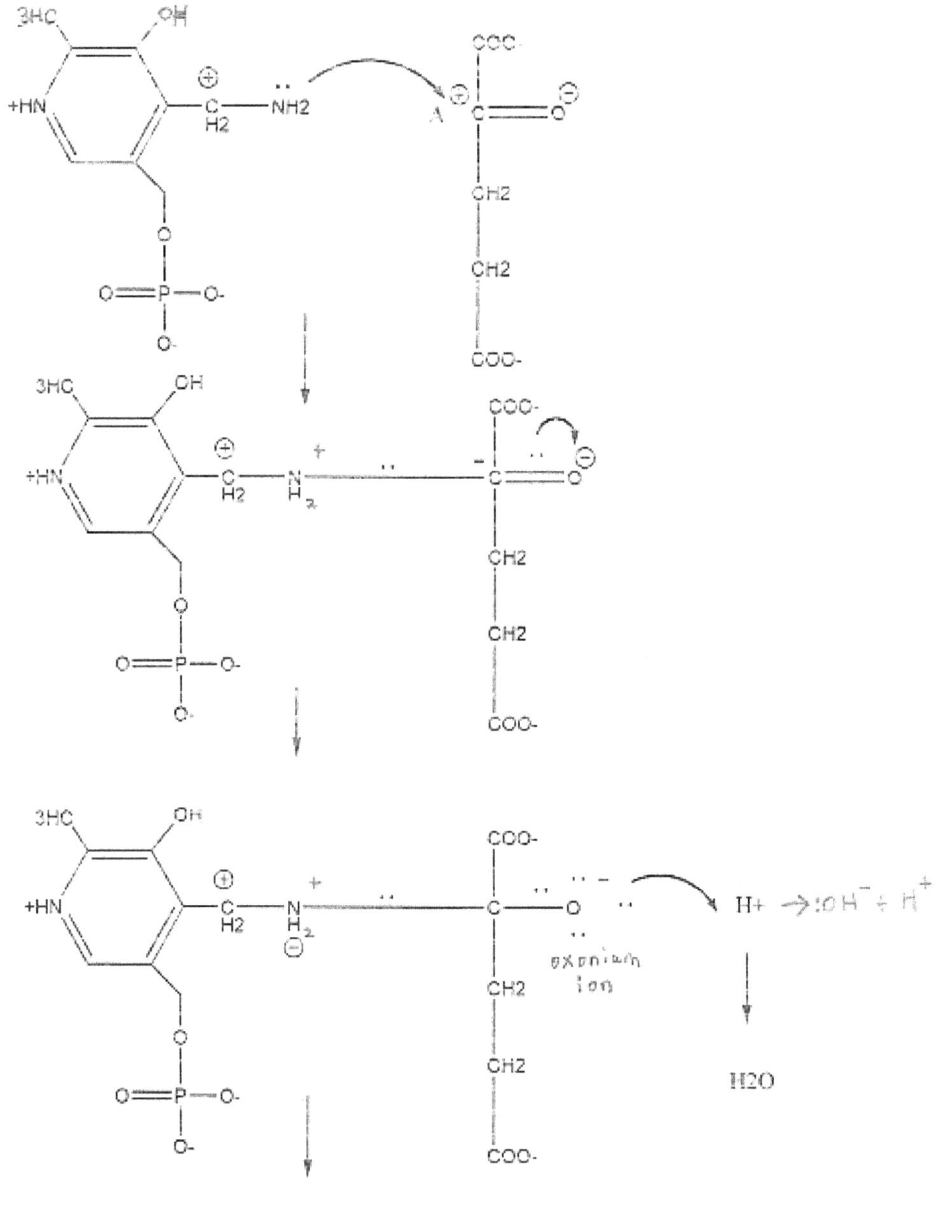
3HC
OH
+HN
H2
NH2
A
COO-
CH2
O
O-
H+
H2O

CARBOCATION

L-Glutamate

H2O

PLP

Notes:

1st Bypass rx. of Gluconeogenesis

In Mitochondria, Pyruvate Carboxyl ase
Addition of CO2

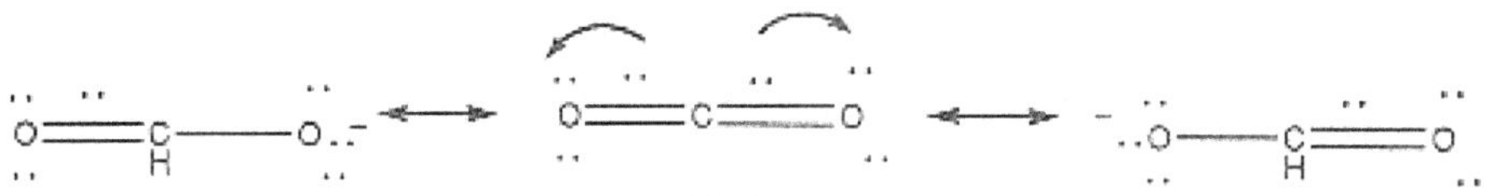

$$O=C-O \cdot \cdot ^{-} \quad \longleftrightarrow \quad O=C=O \quad \longleftrightarrow \quad ^{-} \cdot \cdot O-C=O$$

CO2

Resonance

Overall rx.

2 Pryuvate + 2 HCO3- + 2 ATP -> 2 Oxaloacetate + ADP + Pi
2 3C   2 C   2 4C

Reactants

ATP   PDA—O—P(+)—O $^{-}$

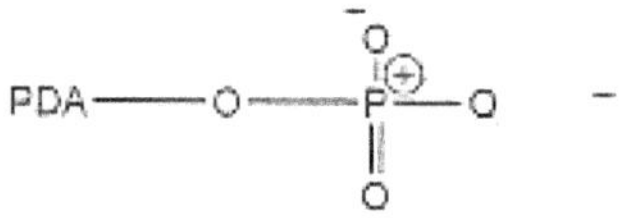

Biotin-Enz.

Lysine   Amide
epsilon-amino grp.

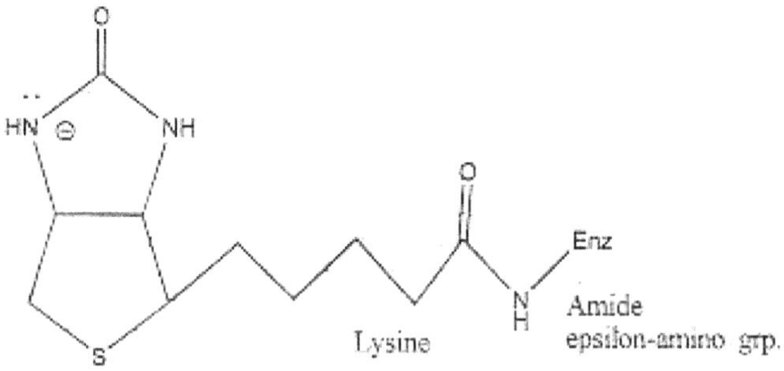

Bicarbonate   HCO3-

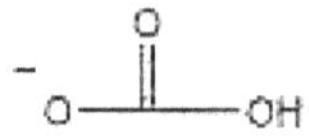

Pyruvate   CH3 / O=...—COO-

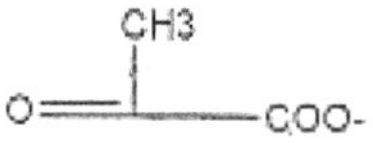

At Catalytic site 1

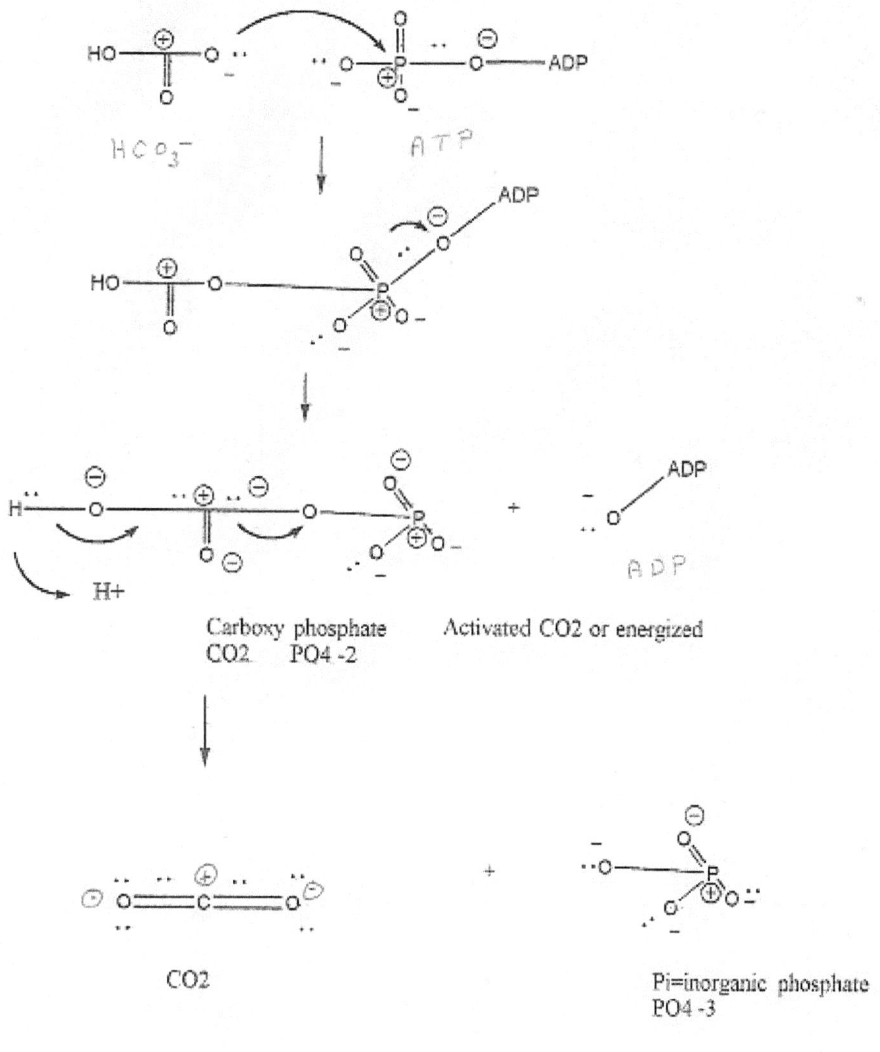
HCO3−
ATP
ADP
H+
ADP
Carboxy phosphate
CO2 PO4 -2
Activated CO2 or energized
ADP
CO2
Pi=inorganic phosphate
PO4 -3

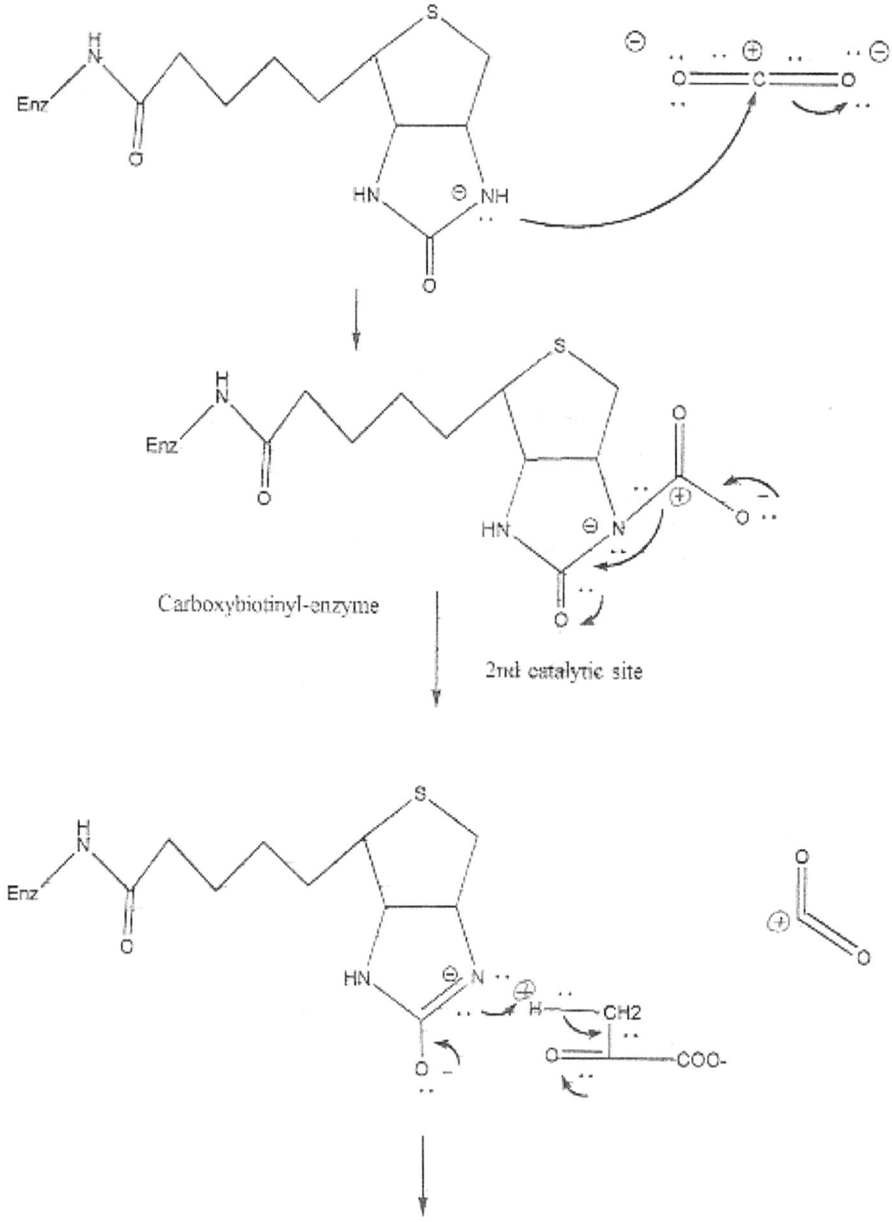
Enz
S
HN
NH
O
Carboxybiotinyl-enzyme
2nd catalytic site
CH2
COO-

Pyruvate enolate

Oxaloacetate

The enzyme has flexible arms to carry the intermediates between catalytic sites.

The double bond can act as a nucleophile.

In the cytosol of the mitochondria, the oxaloacetate is converted to Malate for transport across the inner membrane. This is performed by the transporter protein called Malate-alpha-ketogluterate transporter. Now in the matrix of the mitochondria, Malate is converted back to oxaloacetate. These reactions are reduction type using NADH + H+ as the agent by the enzyme Malate dehydrogenase. This is part of a general scheme used in liver, kidney and heart to transport reducing equivalents between cytosol and matrix in the mitochondria. This is named the Malate-Aspartate shuttle.
We have seen this reduction reaction prior in the fatty acid metabolism section above.

Oxaloacetate + (NADH+) + (H+) → Malate
                              ←

(NADH+) + (H+) : a reversible electron pair sink + 1 proton (H: -) Hydride ion.
H:- → (H+) + (:) Hydride ion is very reactive, unstable.

Oxaloacetate: -OOC-CH2-CO-COO-
                  1   A   B
Can you draw the reaction mechanism for this reaction?

Hints: First draw the structure of Oxaloacetate.
           C labeled B is the reaction site.
           Reducing equivalents add to this C.
           How?

1 possible answer: 1. oxonium intermediate. The radicalization of oxygen.
                       2. Carbo-cation intermediate. The addition of Hydride ion.
With these hints, now draw the reaction mechanism. Enjoy the organic chemistry in action.

Now let us review the second reaction of the first bypass step of gluconeogenesis.
There are 3 actors in this reaction:

1. Oxaloacetate
2. GTP
3. PEP Carboxy kinase

This is a classic nucleophilic reaction mechanism with the intermediate of a carbocation.

The original CO_2 fixation of the immediate prior reaction is removed in the process, in other words, reversed. The product is PEP which is an intermediate of glycolysis. The 5 reaction steps that follow are reversible reactions of glycolysis. This is energy and material conservation at work. No new materials are required to the product DHAP, Dihydroxyacetone phosphate.

As an exercise, try to write the reaction mechanisms of the 5 reactions. When you have an answer, e-mail me to collaborate.

1. PEP $\rightarrow$
2. 2-PG $\rightarrow$
3. 3-PG $\rightarrow$
4. 1,3-BPG $\rightarrow$
5. DHAP

From DHAP $\rightarrow$ F 1,6-BP is the second bypass reaction of gluconeogenesis. This reaction requires F 1,6-BPase.

Afterward, the rest of gluconeogenesis is the following:

1. F 1,6-BP $\rightarrow$
2. F 6 P $\rightarrow$
3. G 6 P $\rightarrow$ requires G 6 Pase The third bypass step in gluconeogenesis.
4. Glucose

Once at this stage, Glucose passes through glycolysis again to the pyruvate product which is converted to acetyl CO-A for entry into the Citric Acid Cycle. From here, compounds proceed to the oxidation phosphorylation chain to harvest the energy transfer which creates ATP. ATP provides the energy necessary for all cell functions. The matter is converted to CO_2 + H_2O, complete combustion. This is why glucneogenesis is the long way to fast energy production. This is the route of last resort. This is not the preferred route for acute energy needs. Glycogenolysis, simple sugars, and fats are.

So, prior to a football game, this is why the carbo-loading regimen is necessary.

Let us get back to the second reaction of the first bypass step.

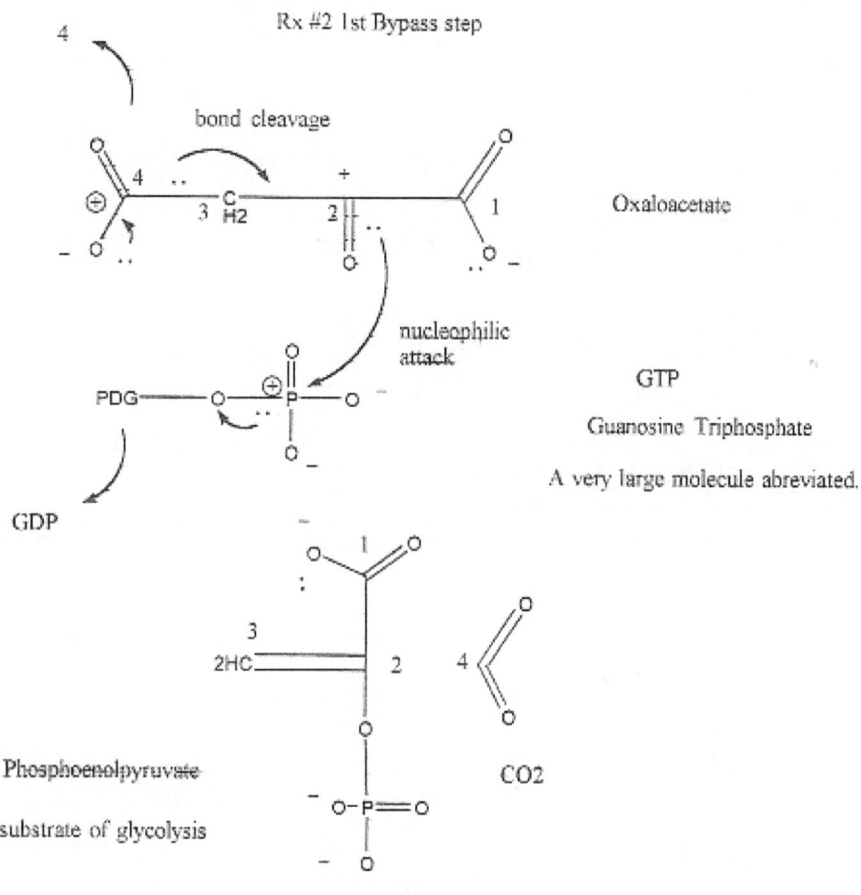

Rx #2 1st Bypass step

bond cleavage

Oxaloacetate

nucleophilic attack

GTP

Guanosine Triphosphate

A very large molecule abreviated.

GDP

Phosphoenolpyruvate

substrate of glycolysis

CO2

PEP Carboxy kinase
C=O   +PO4 -2

Let us do the reaction steps individually.

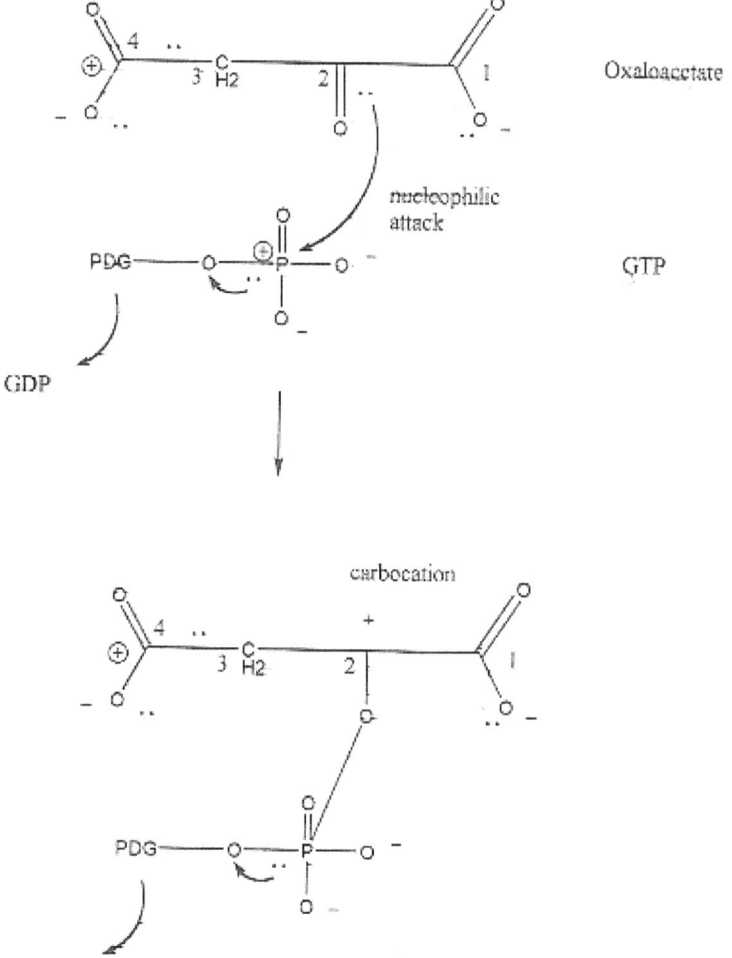

Oxaloacetate

GTP

GDP

carbocation

GDP

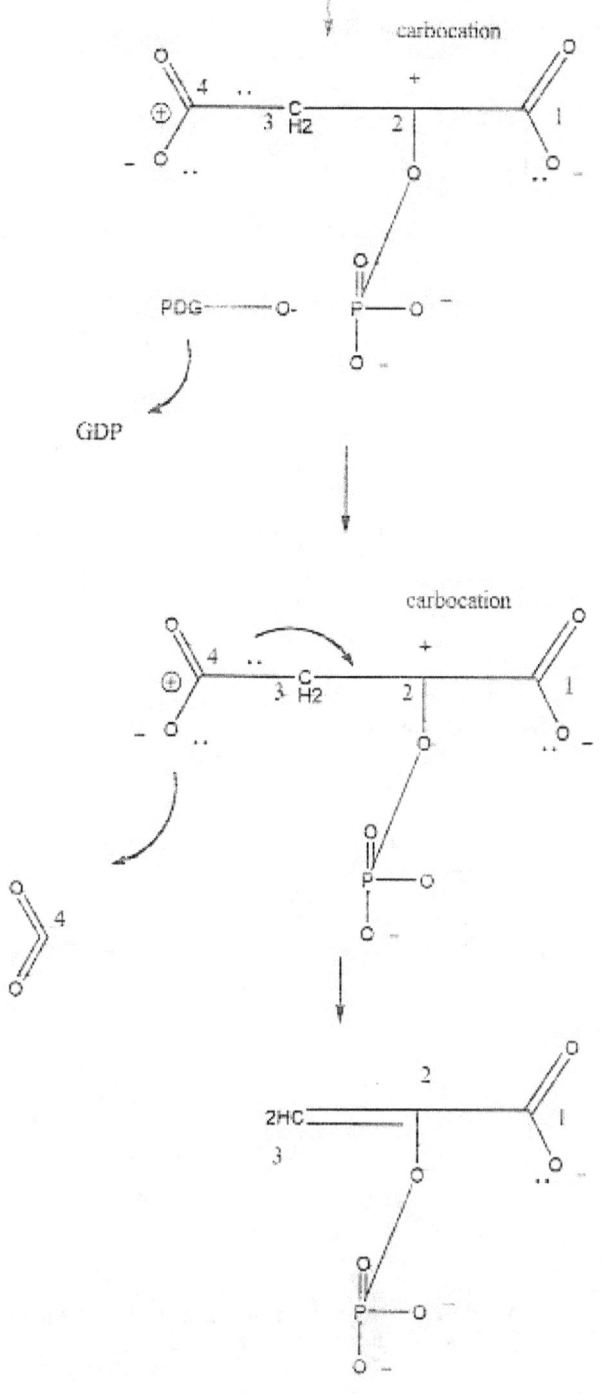

In the Liver, there is an alternative pathway to PEP.
Lactate built from vigorous exercise, under anaerobic conditions, is converted to
pyruvate via the Cori Cyle.

In the cytosol of the hepatocyte, lactate -> pyruvate: by Lactate Dehydrogenase

In the mitochondria, pyruvate -> oxaloacetate -> PEP: by pyruvate caroxylase + PEP carboxykinase
PEP enters the glycolytic pathway in reverse.

Lactate

Pyruvate

NAD+ is an oxidizing agent which gets reduced.

Lactate is a reduced pyruvate which gets oxidized.

With Lactate drawen like this, can you see a mechanism for
oxidation, reduction?
The oxidation occurs at the alcohol function.
Polarity, acidity, and Oxidation-Reduction is at work in this
reaction.

So propose a reaction mechanism! Draw the intermediates!

Hydride ion Addition: H:- This is equivalent to an
electron pair transfer.
See (NAD+) + (H+) reaction mechanism.

Remember covalent bonds are electron pairs which can react accordingly

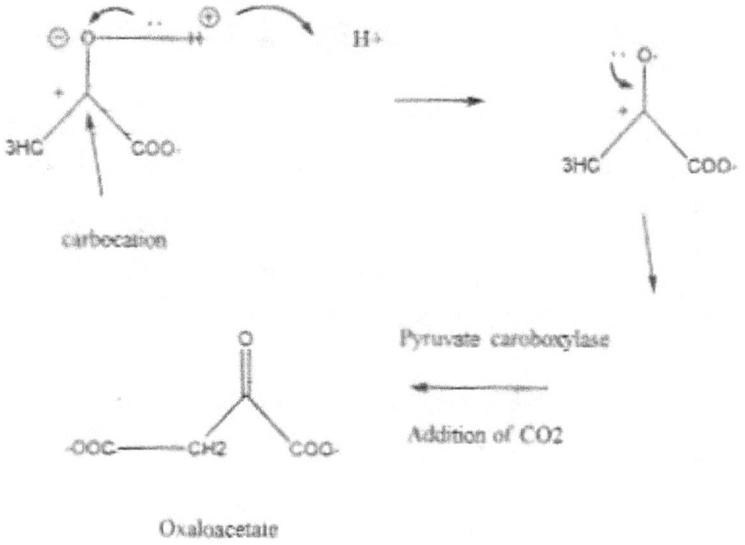
H+
3HC
COO-
carbocation
Pyruvate caroboxylase
Addition of CO2
-OOC
CH2
Oxaloacetate

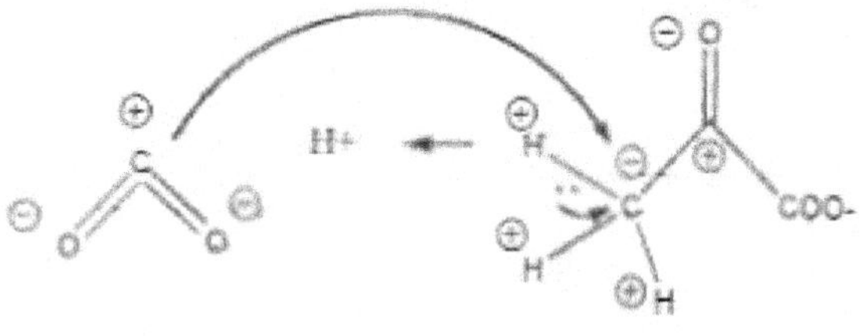
H+
COO-

Methy group of pyruvate is polarized and turns acidic which increases the probability of CO2 Additon.

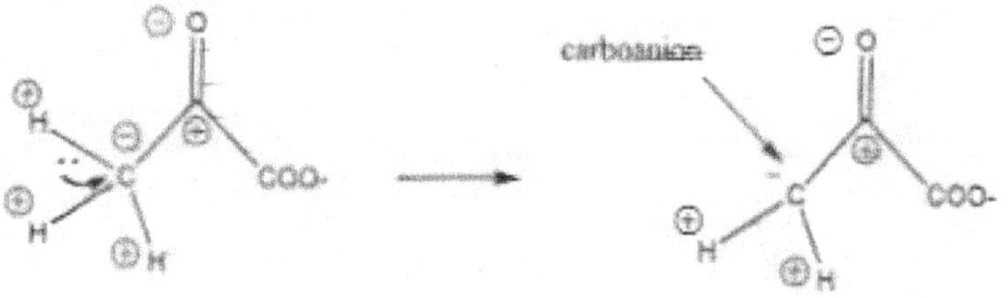
COO-
carboanion

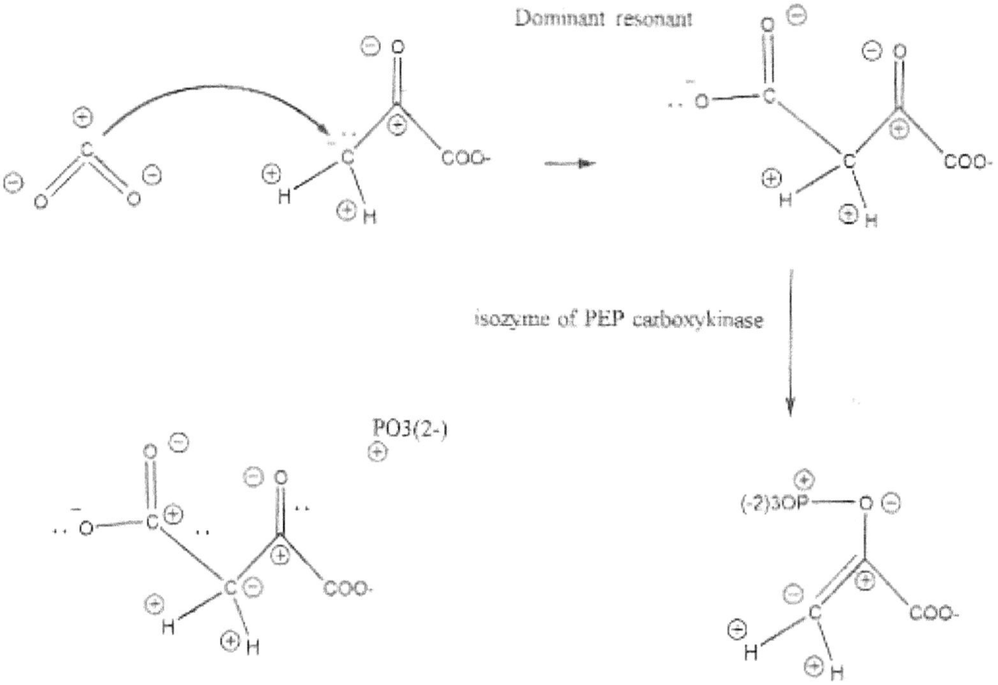
Dominant resonant
isozyme of PEP carboxykinase
PO3(2-)
(-2)3OP
COO-

You have seen the mechanism before.
Can you draw the curly arrow notation?
Draw the intermediates!

PEP

The methylene group of OA turns more neutral or positive relative to its already negative polarity. Therefore the covalent electron bond pair favors this C atom creating a double bond in PEP.

Use this space for your work.

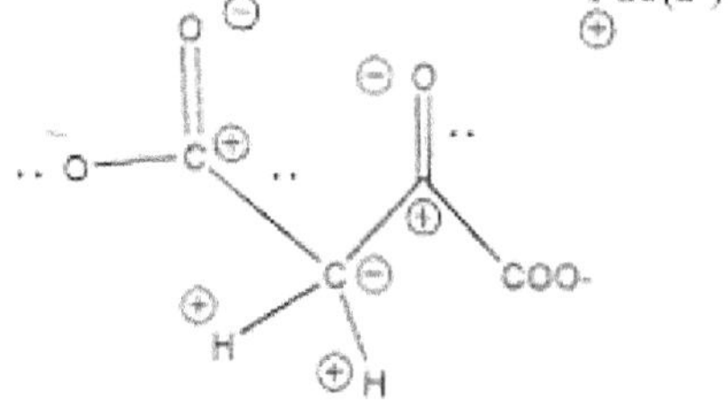
PO3(2-)
COO-

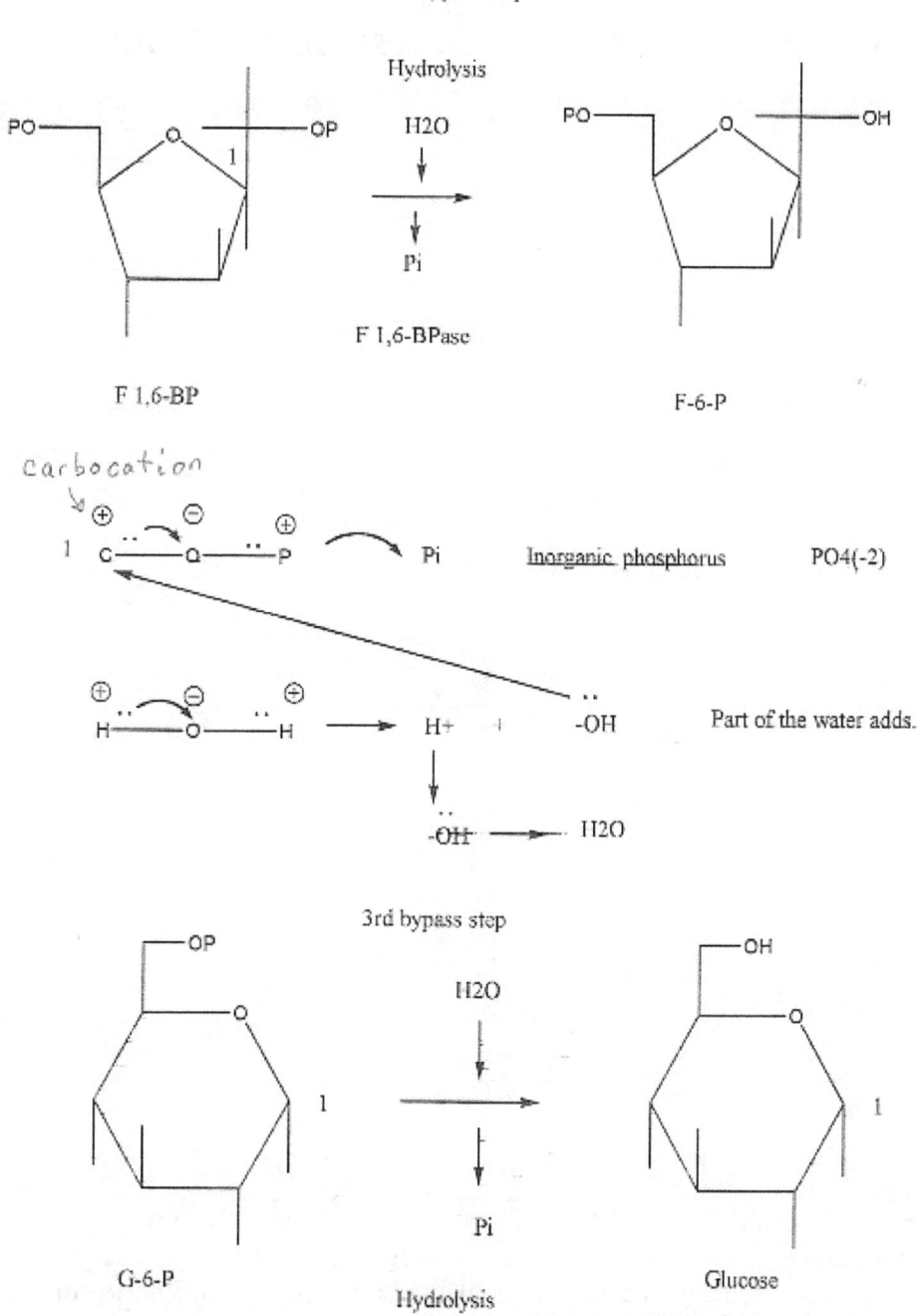

2nd bypass step

Hydrolysis

H2O

Pi

F 1,6-BPase

F 1,6-BP

F-6-P

carbocation

1 C — O — P → Pi   Inorganic phosphorus   PO4(-2)

H — O — H → H+ + -OH   Part of the water adds.

-OH → H2O

3rd bypass step

H2O

Pi

G-6-P

Glucose

Hydrolysis

G-6-Pase   found in liver/kidney; not muscle/brain

Cory

Chapter 10

The Generation of CO2: The First half of Combustion

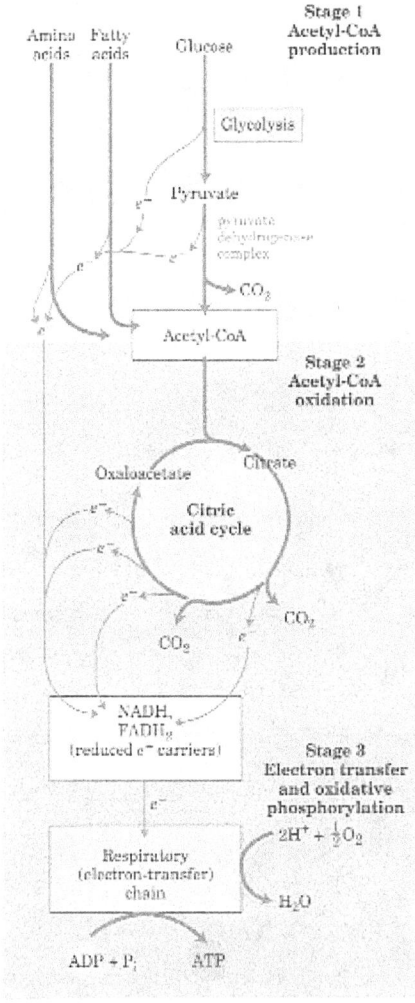

In any scientific laboratory, complete combustion leads to CO2 and H2O.

Were are at the center of the energy process of biological systems.

In the first part of the book, we dealt with the degradation of nutrients to the point of pyruvate and acetyl CoA. This is the first third of the above figure, Stage 1 Acetyl-CoA production. You can see the 3 nutrient groups are brought to the Acetyl-Co A level.

Carbohydrates → pyruvate → Acetyl-Co A

Fats → Acetyl-CoA

---

1   Nelson, David L.,Cox, Michael M.,**Lehninger, Principles of Biochemistry,** 4th Edition, p.602.

Proteins → aa → Acetyl-CoA

As noted in the figure, some amino acids can transform into Acetyl-CoA directly.

There is something very important depicted in the figure; and that is the generation of electrons. These carry the energy of chemical transformation, chemical potential energy. This energy can be used for all energy consuming biological processes. The chemical currency of the system is ATP. This is by far the most important form of chemical potential energy.

This chapter will cover the reactions of the Citrus Acid Cycle. This generates CO_2 and electrons.

The Respiratory chain will be covered in the next chapter. This is where ATP and H_2O is generated. Water being the other half of complete combustion.

This is life's way of combustion. Biochemists call this cellular respiration.

Before we start with the TCA cycle, we need to look at pyruvate → Acetyl-CoA first.

The combustion reactants are either at Acetyl-CoA or pyruvate stage. To enter the TCA cycle, all have to be at the Acetyl-CoA stage.

Notes:

Pyruvate Dehydrogenase Complex (PDH Complex E1,E2,E3)

Oxidative Decarboxylation

CoASH=Pantothenate=B6 pantothenic Acid
NAD+ Nicotinoamide=B5 Niacin
TPP=Thiamine=B1
FAD=Riboflavin=B2
Vitamin Bs are important in carbohydrate metabolism. This is the reason for suppling these vitamins during the carbo-loading process.

3 enzymes 5 reactions: E1 Pyruvate dehydrogenase reactions 1,2
                                                    - H
                           E2 Dihydrolipoyl transacetylase: reaction 3
                                                    +/- CH3-C=O-
                           E3 Dihydrolipoyl dehydrogenase: reactions 4,5
                                                    - H

The enzyme complex demonstrates the concept of substrate channeling. The enzyme holds on to the substrate during the 5 reactions until the final product is reached. This enables the reaction to be proficient and accurate. Spending the least amount of energy and time to complete the reaction sequence. It is truely amazing how this protein complex came to existence. Considering the thermodynamics involved. What was involved in the natural selection of this complex? It had to conform to the laws of conservation of matter and energy. These reactions were probably separate during earlier times. Evolution brought these reactions together. The major influencing factor was the creation of the enzyme itself. In otherwords, the evolution of the complex itself. Biomolecules follow the laws of physics. Looking at it's final position in metabolism, this complex plays a crucial role to the survival of the cell.

The individual structures will be showen during each indivdual reaction.
The structures will be abbreviated for simplicity and clairity.

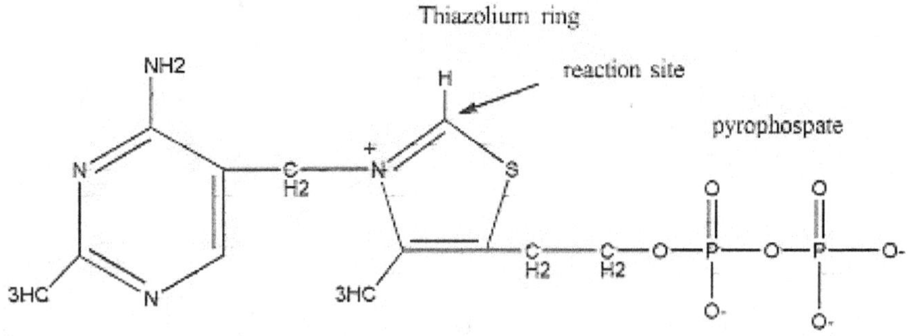

Thiazolium ring

reaction site

pyrophospate

Thiamine pyrophosphate
B1

Because of the conjugate double bonds, resonance is possible.
Note the N+ in the ring. This can act as an electron sink site. Further increasing the resonant character of the reaction.
The reaction ring will be enlarged for clarity.

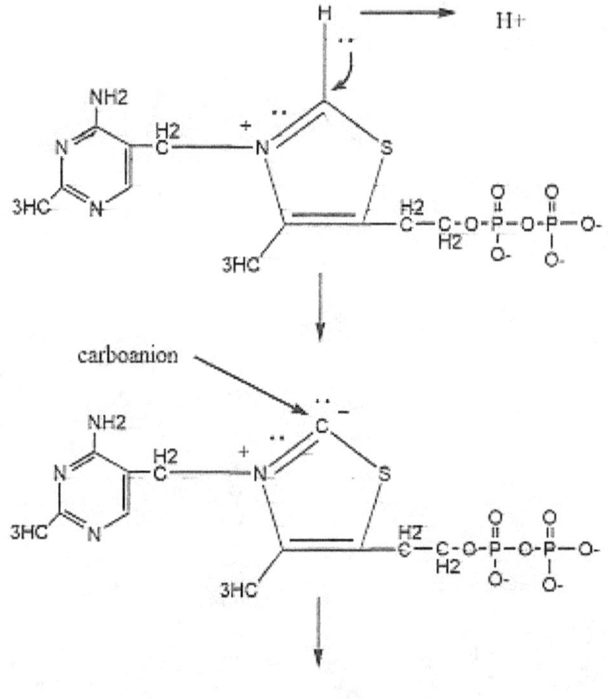

H+

carboanion

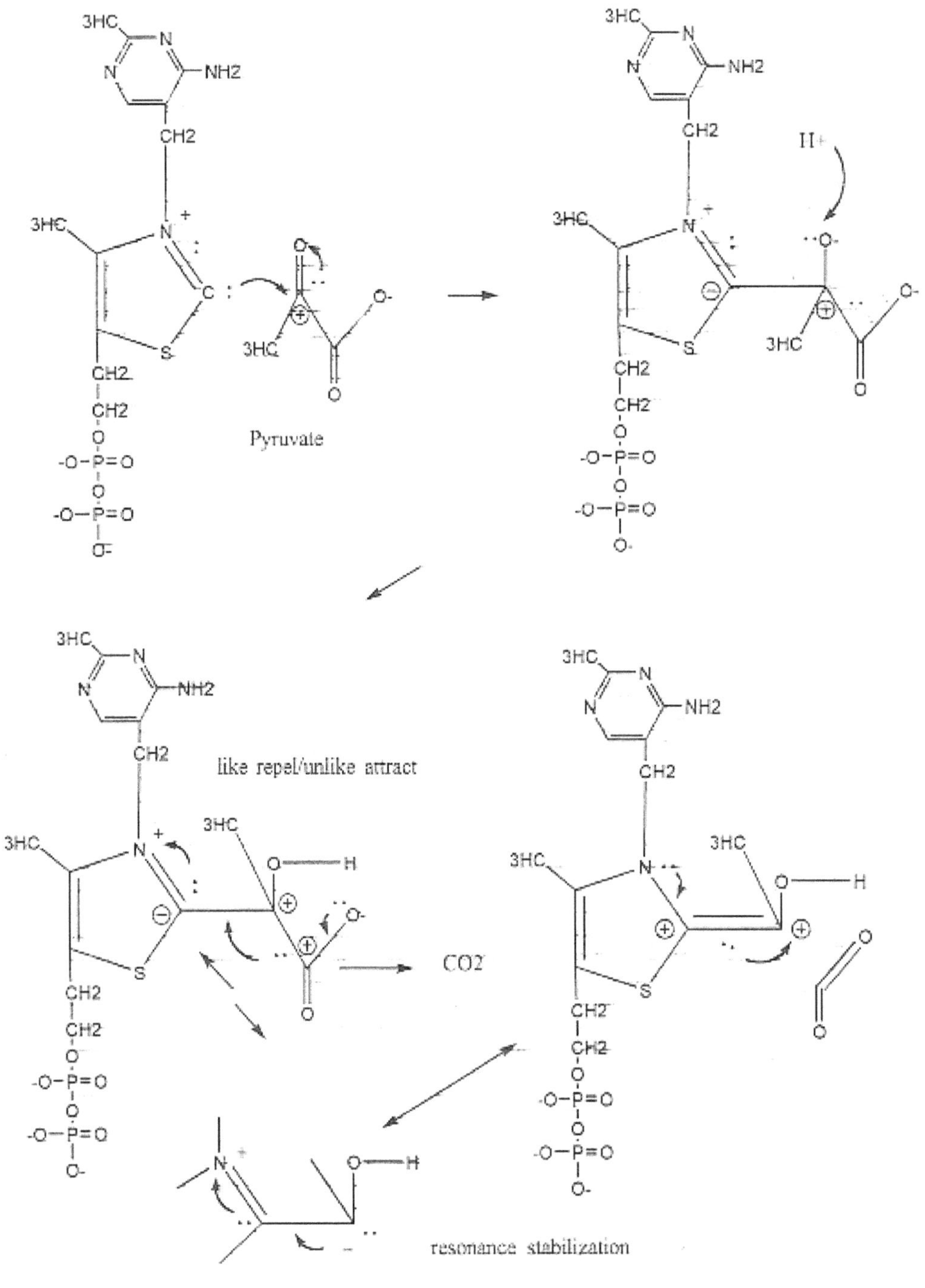

Pyruvate

like repel/unlike attract

CO2

resonance stabilization

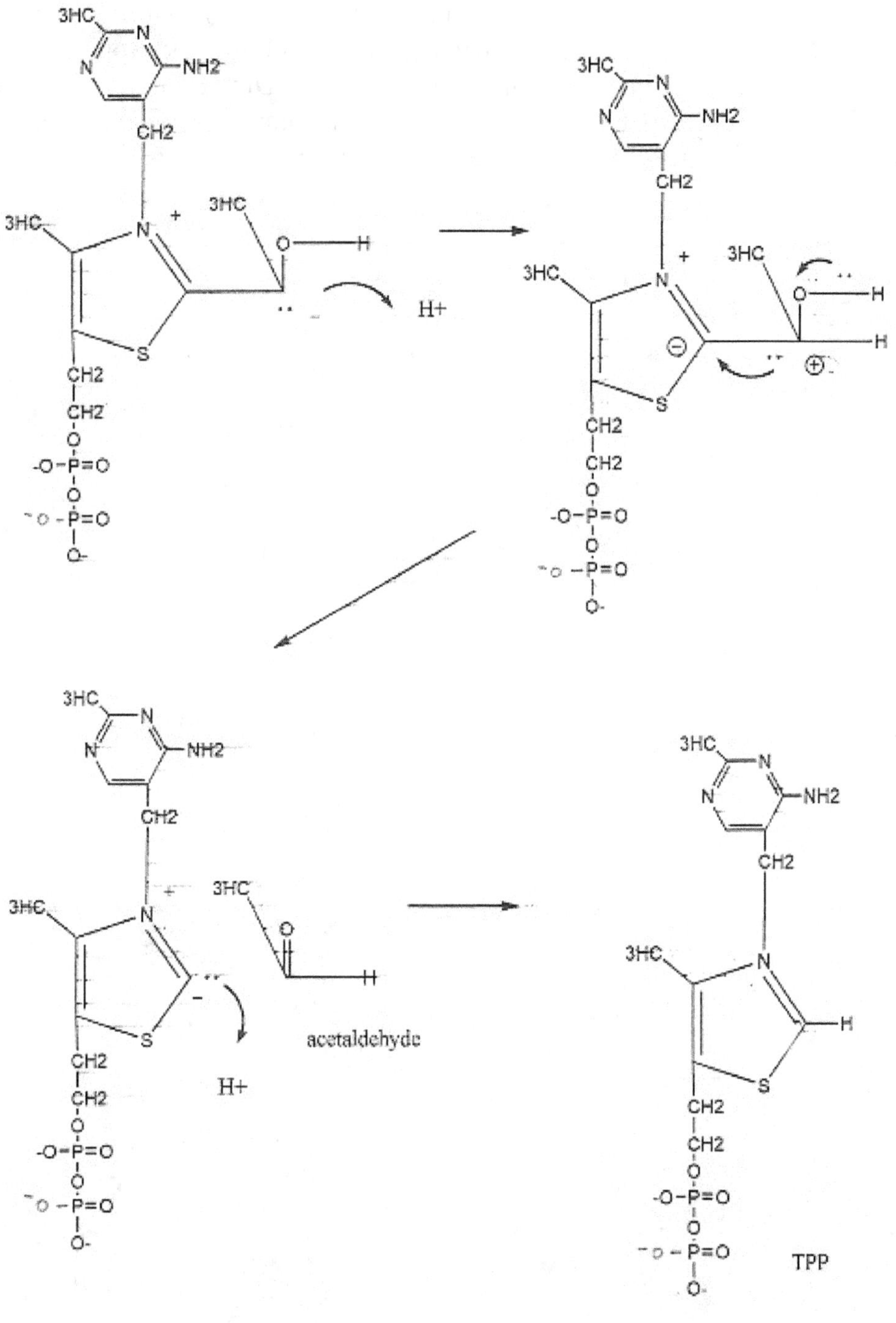

acetaldehyde

H+

TPP

Remember, substrate channeling is in effect. The reactants and, more importantly, the products remain attached to the coenzyme. TPP, the coenzyme, stays attached to E1, the enzyme, Pyruvate dehydrogenase. The above reaction continues to the hydroxyethyl derivative, the structure before acetaldehyde release. The above reaction is step #1 and is called decarboxylation, removal of CO2. Reaction step #2 is done by the same enzyme. It includes the transfer of 2 electrons and the acetyl group to form the acetyl thioester. This will require lipoate and E2, Dihydrolipoyl transacetylase. What is not shown here is where does E1 attach to TPP? I see an amide linkage between the NH2 of TPP and the COO- group of Aspartate or Glutamate of the polypeptide chain of E1.

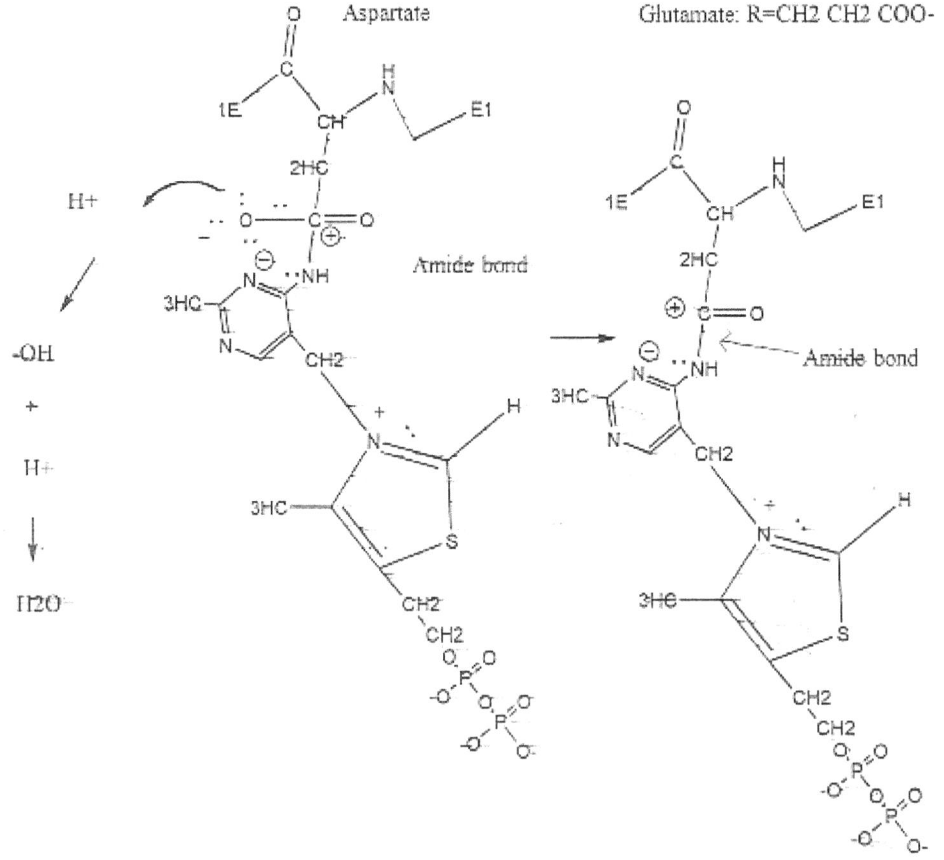

oxidized: electron rich

Step #2 E1

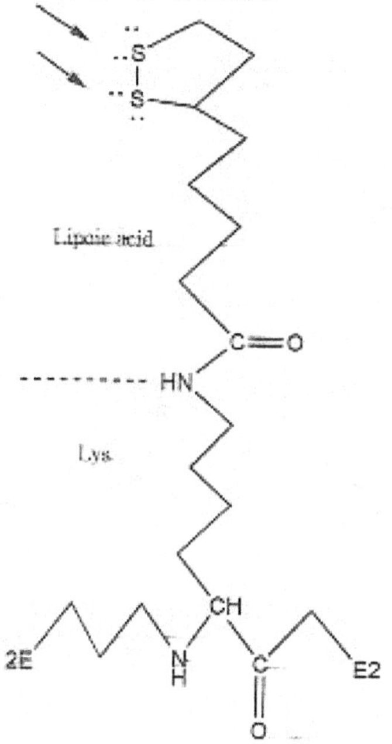

lipoyllysyl moiety=prosthetic group of E2
It carries H and acetyl/acyl group.

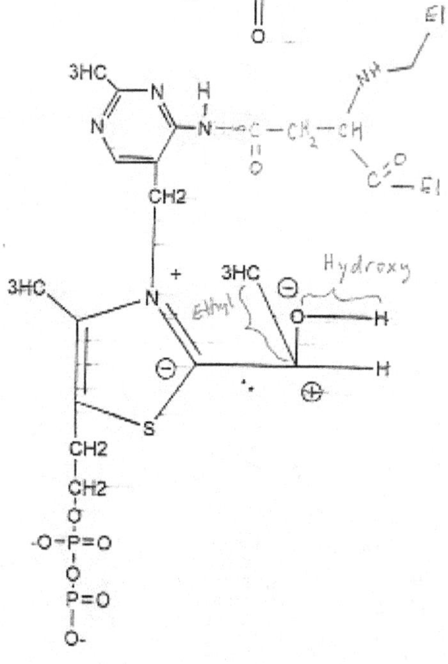

Lipoic acid

HN

Lys.

2E
NH
O
E2

E1

3HC
N
N
H
N—C—CH2—CH
NH
O
C=O—E1

CH2

3HC
N
+
3HC
Hydroxy
Ethyl
O—H
⊖
⊖
••
⊕
H
S

CH2
CH2
O
-O-P=O
O
P=O
O-

Hydroxyethyl-TPP

S
S

HN
C=O

2E
NH
CH
E2
O

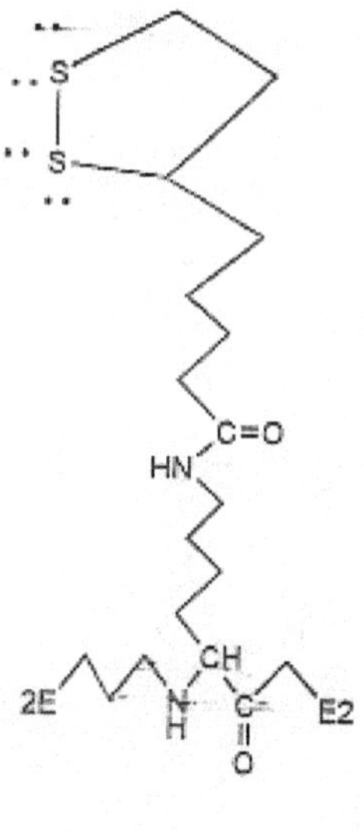

lipoate

As you can see the product of the first step is still attached to E1 and now becomes reactant #1 in step #2. The second reactant in step 2 is attached to E2. E1 performs the reaction.

A double de-protonation.

transfer of 2 electrons +
Acetyl group

Can you identify the above transfer?

Acetyl   electron pair
group   as covalent bond

Can you draw the individual reaction
mechanisms?
Use the above curely arrows as a guide.

Step #3 by E2

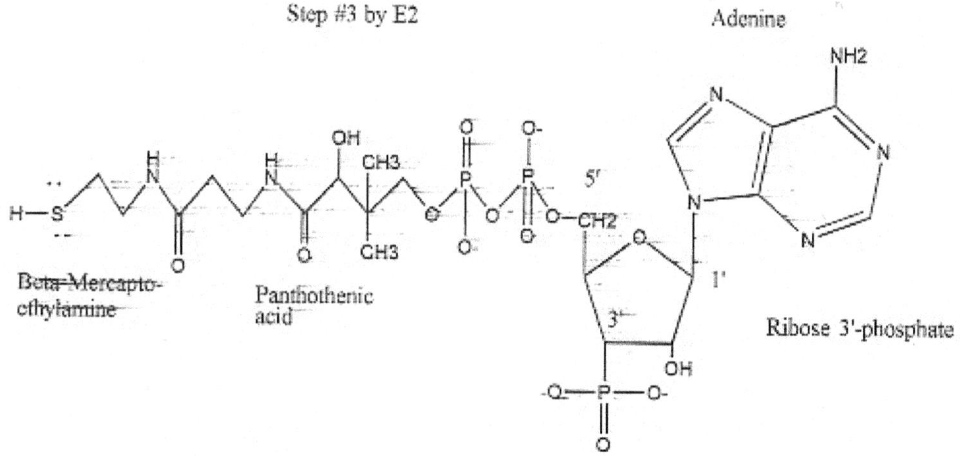

Adenine

NH2

OH
CH3

5'

CH3

H—S

N
H

N
H

O

O

O

P

O

P

O

O

O⁻

O⁻

O—CH2

O

1'

3'

OH

Beta-Mercapto-
ethylamine

Panthothenic
acid

Ribose 3'-phosphate

O⁻—P—O⁻

O

3'-Phosphoadenosine diphosphate

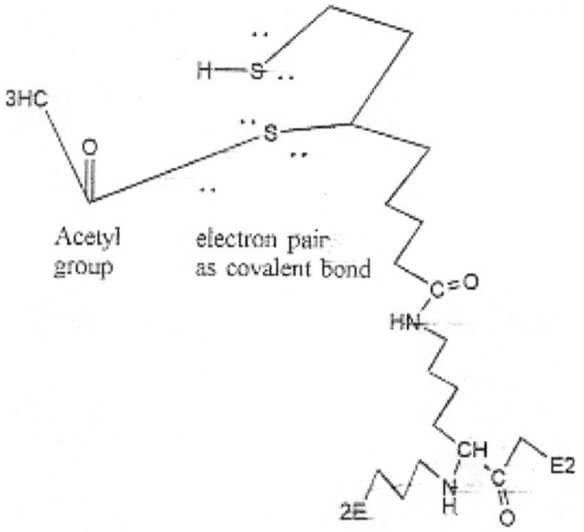

3HC

H—S

S

O

Acetyl
group

electron pair
as covalent bond

C=O

HN

CH

E2

2E

N
H

O

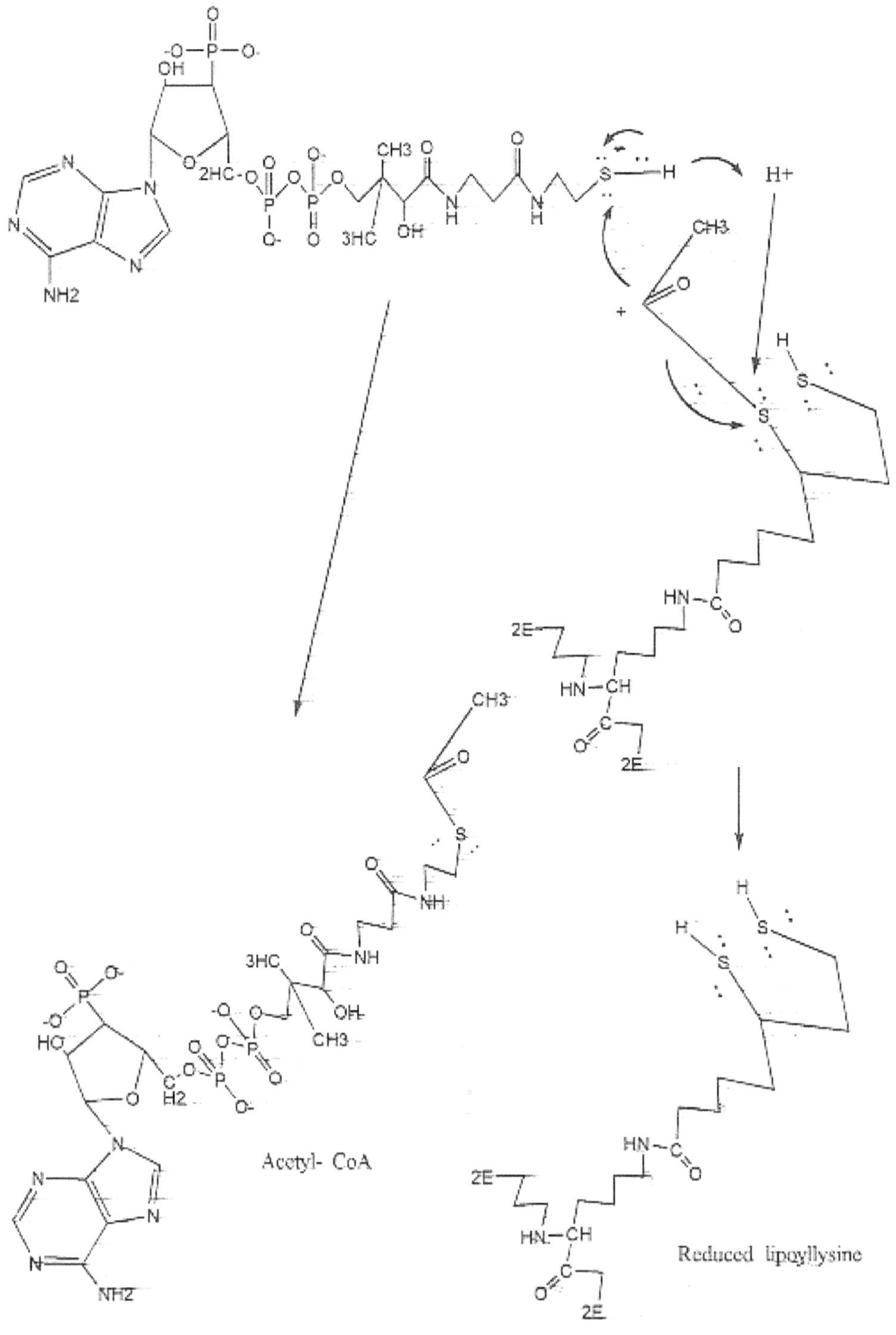
H+
Acetyl- CoA
Reduced lipoyllysine

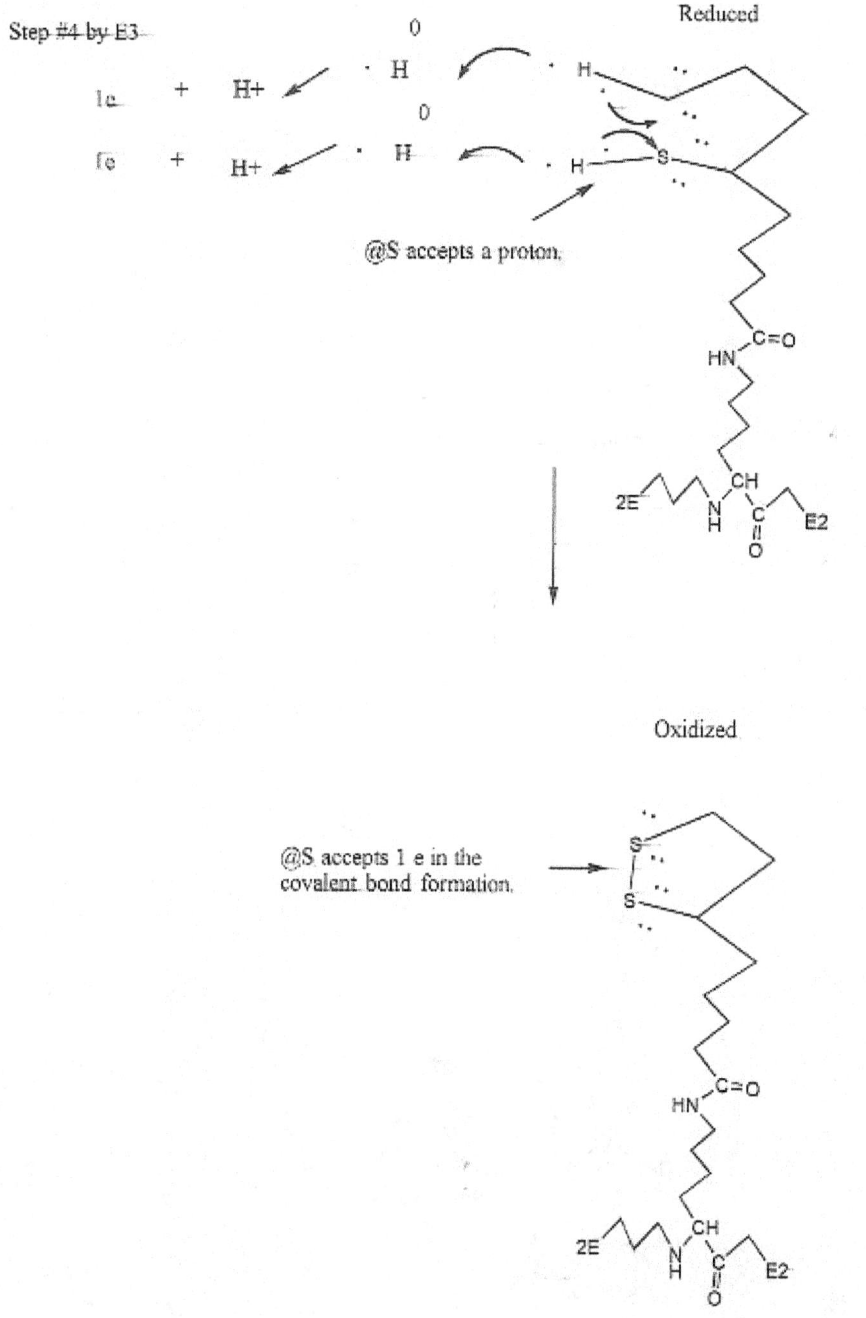

Step #4 by E3

Reduced

1e   +   H+   ←   · H
                    0

fe   +   H+   ←   · H

@S accepts a proton.

HN—C=O

2E   N   CH   E2
       H   C
           O

Oxidized

@S accepts 1 e in the
covalent bond formation.   →

HN—C=O

2E   N   CH   E2
       H   C
           O

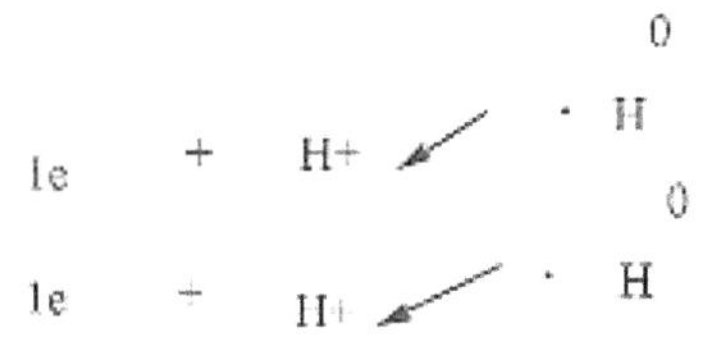

$$1e \quad + \quad H+ \quad \longleftarrow \quad \cdot H \quad \overset{0}{}$$

$$1e \quad - \quad II+ \quad \longleftarrow \quad \cdot H \quad \overset{0}{}$$

FAD

Isoalloxazine ring

Resonant ring : conjugation area

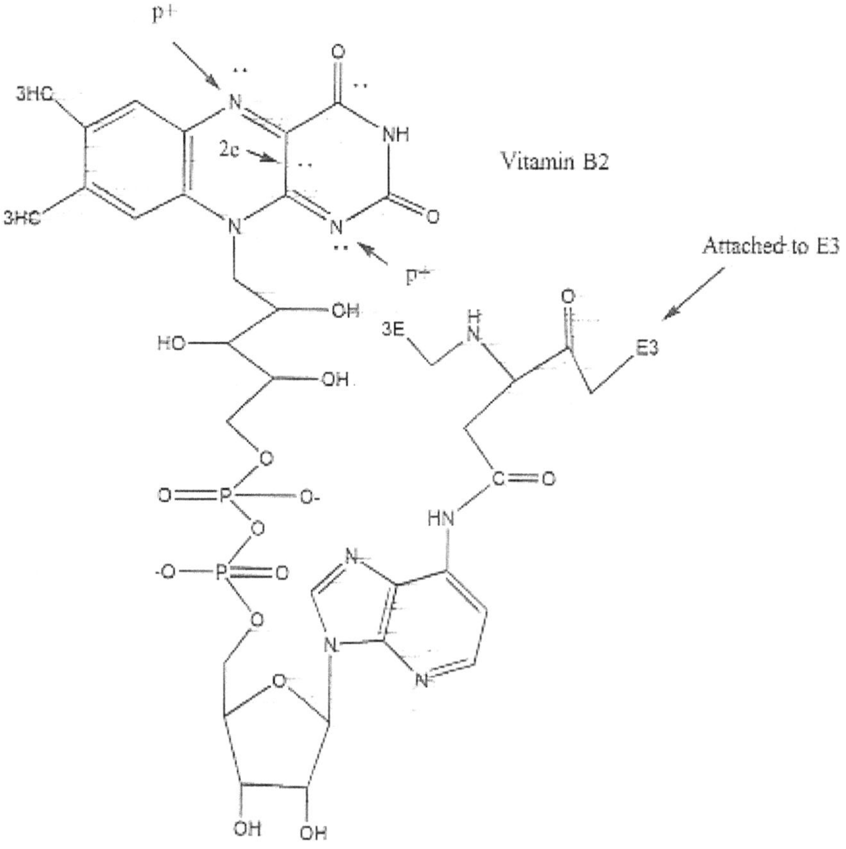

p+

3HC

2e

3HC

Vitamin B2

P+

Attached to E3

3E

H
N

E3

OH

HO

OH

O

P

O-

O

-O-

P

O

O

HN

O

N

N

N

N

OH   OH

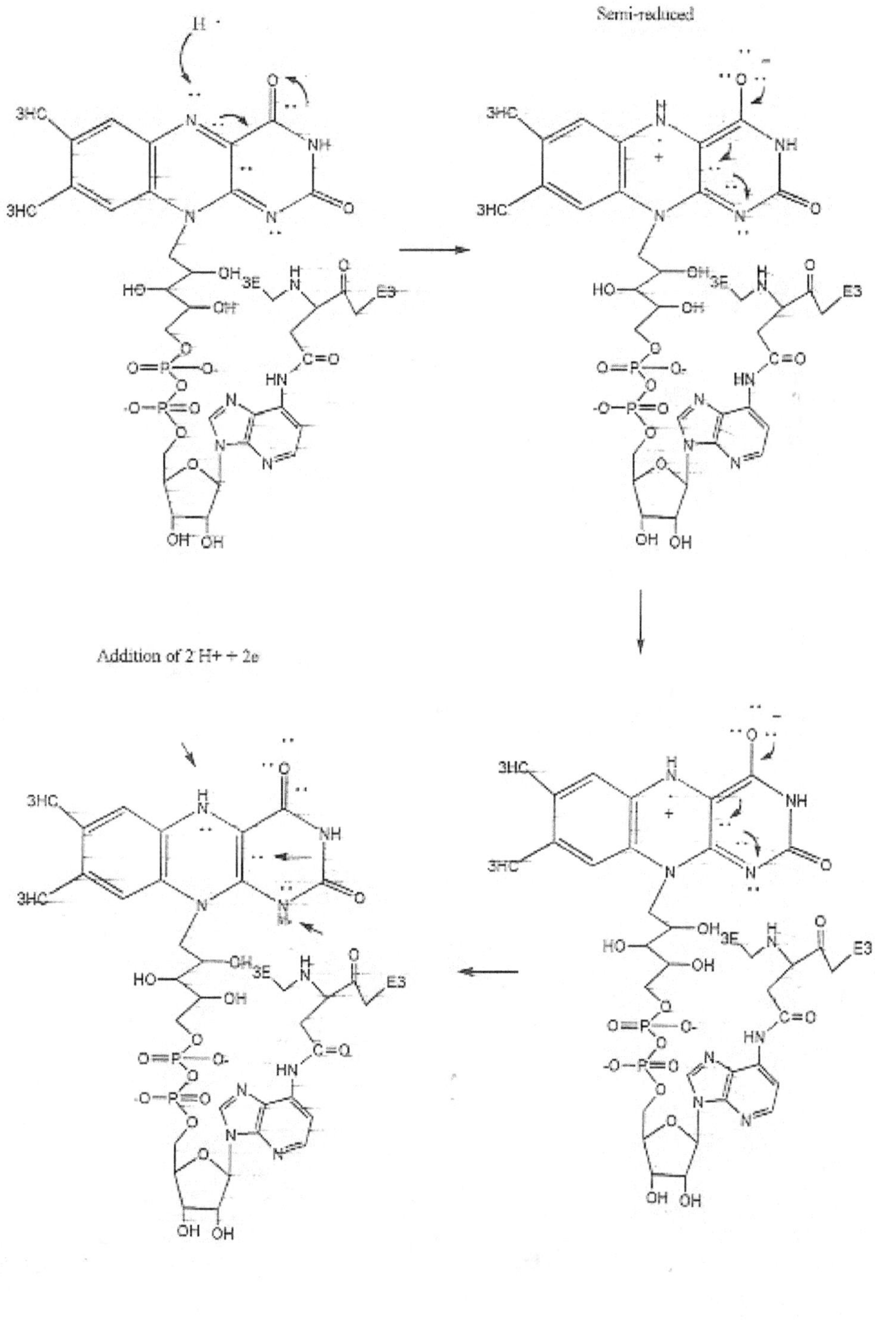

Semi-reduced

Addition of 2 H+ + 2e

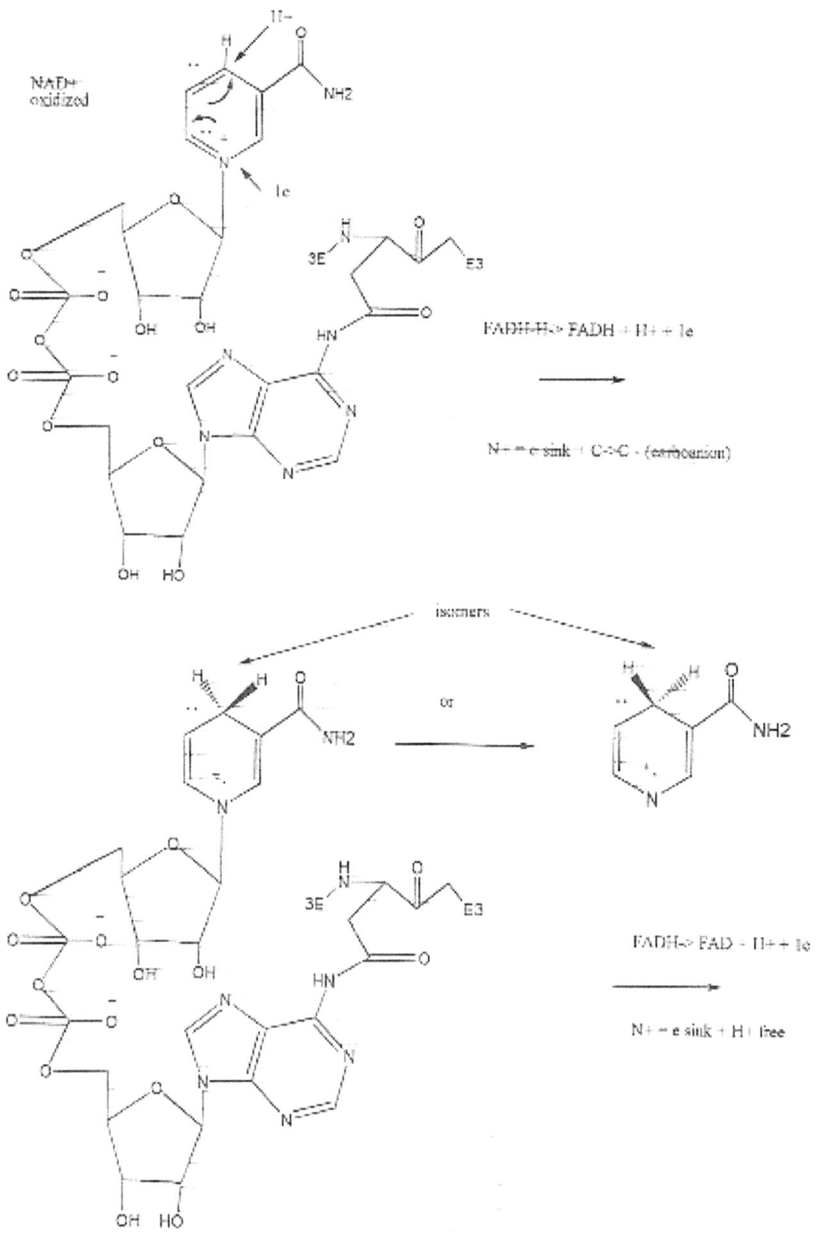

NAD+
oxidized

FADH-H-> FADH + H+ + 1e
⟶

N+ = e sink + C=C - (carbanion)

isomers

or
⟶

FADH-> FAD + H+ + 1e
⟶

N+ = e sink + H+ free

With the creation of Acetyl-CoA (thioester), it can enter the Citric Acid Cycle. Oxaloacetate, already in the cycle, condenses with Acetyl-CoA to make Citrate. This procedes as an ordered bi-substrate mechanism. First OA binds to the enzyme and is held in place by the enzyme's Arginine residues (H-bond). The carboxyl group of Aspartate acts as a base to remove a proton from Acetyl-CoA. The carbonyl group of A-CoA attacks the amine group of Histidine to create an enolate intermediate. The enolate is a strong nucleophile which undergoes a Claisen Condensation with OA creating the Citroyl-CoA, another intermediate. The Hydroxyl group of water, a strong nucleophile, hydrolyzes the thioester bond to release CoA-SH and Citrate.

The enzyme, with it's many functional side groups in the amino acids, creates an induced fit with the substrate. This stabilizes it's position, ensures the reaction by increasing the accuracy and precision, and completion.

A conformational change, morphing the enzyme structure, occurs to overcome the energy barrier; so the reaction can proceed forward to product.

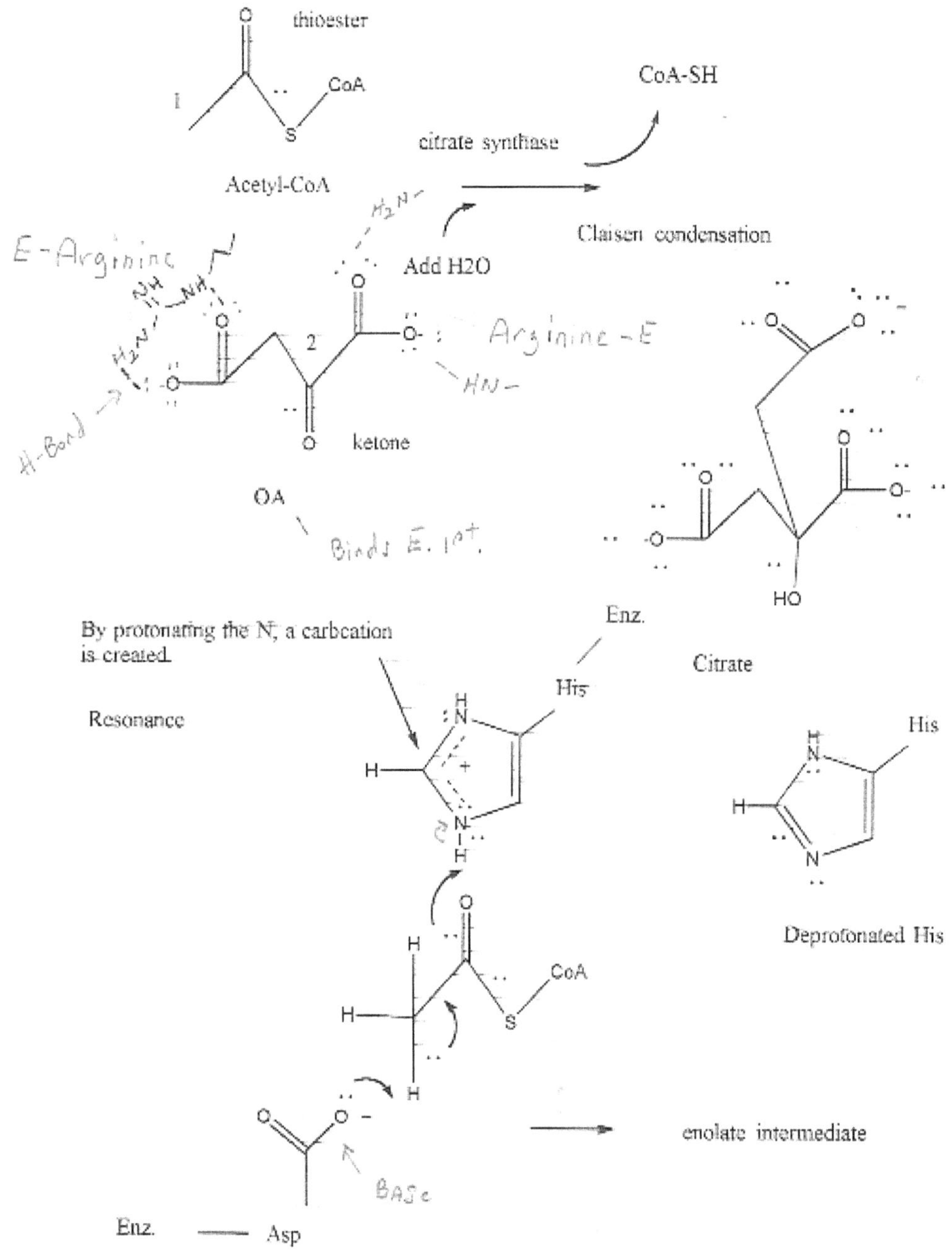

thioester

I

O
‖
CH₃—C—S—CoA ··CoA

Acetyl-CoA

CoA-SH

citrate synthase

Claisen condensation

E-Arginine

H-Bond

Add H2O

Arginine-E
HN—

2⁻
O⁻

ketone

OA

Binds E. int.

Citrate

HO

By protonating the N, a carbcation is created.

Resonance

Enz.

His₅

H
N⁺
H
N
H

Enz.

His

H
N⁺
H

N

Deprotonated His

H
H

O
‖
C—S—CoA ··CoA

H

Enz. ——— Asp

O
‖
C—O⁻

BASe

enolate intermediate

Citroyl-CoA

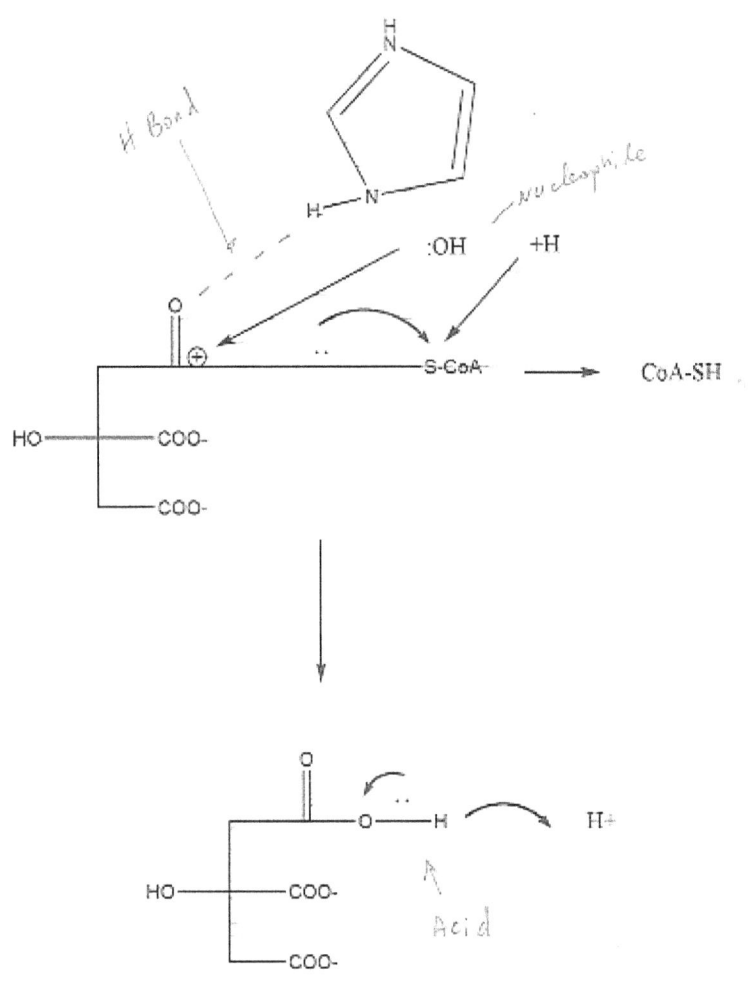

Citrate

O.A. + Enz. → Conformational change . + Acetyl-CoA →
Citroyl-CoA → Conformational change → thioester hydrolysis
"ordered bisubstrate mechanism"

Induced fit

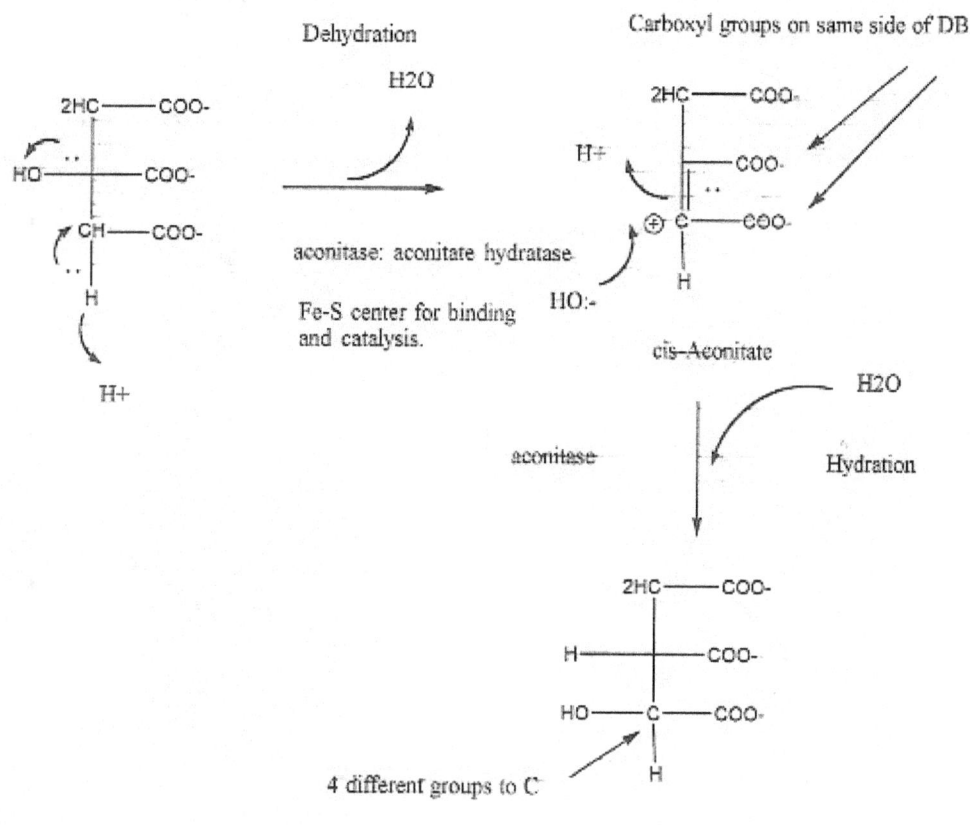

Dehydration

H2O

aconitase: aconitate hydratase

Fe-S center for binding and catalysis.

Carboxyl groups on same side of DB.

cis-Aconitate

aconitase

H2O

Hydration

4 different groups to C

Isocitrate

Iso : means isomer of citrate.

Notes:

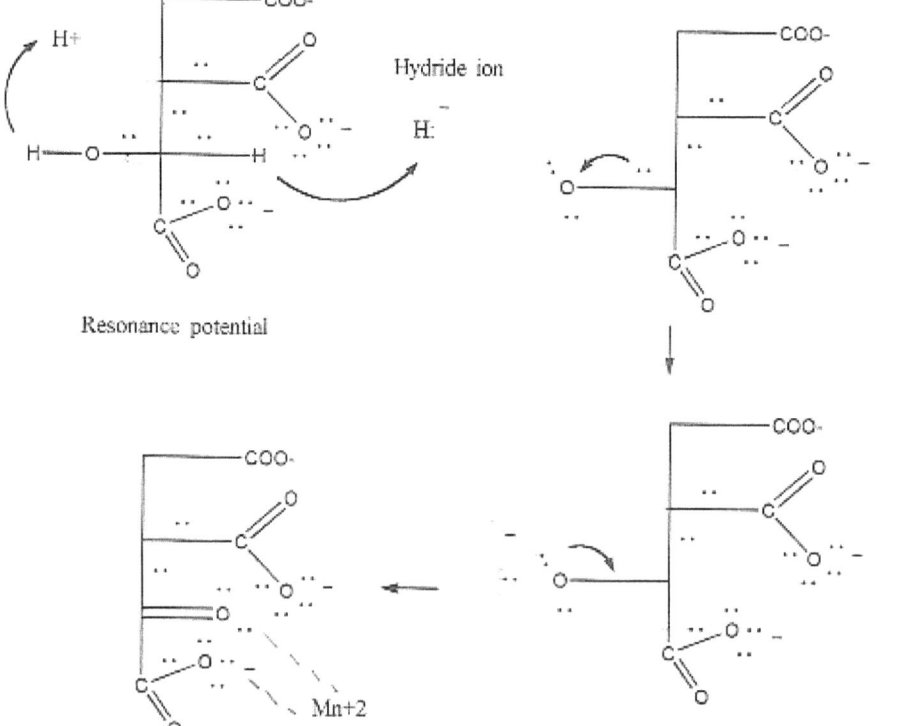

Isocitrate

Isocitrate dehydrogenase
H loss

Oxidative decarboxylation
H+ loss   CO2 loss

Alpha-Ketoglutarate

Resonance potential

Hydride ion

Oxalo succinate

Metallic induction by coordinate bond

Alpha-Ketoglutarate

Notes:

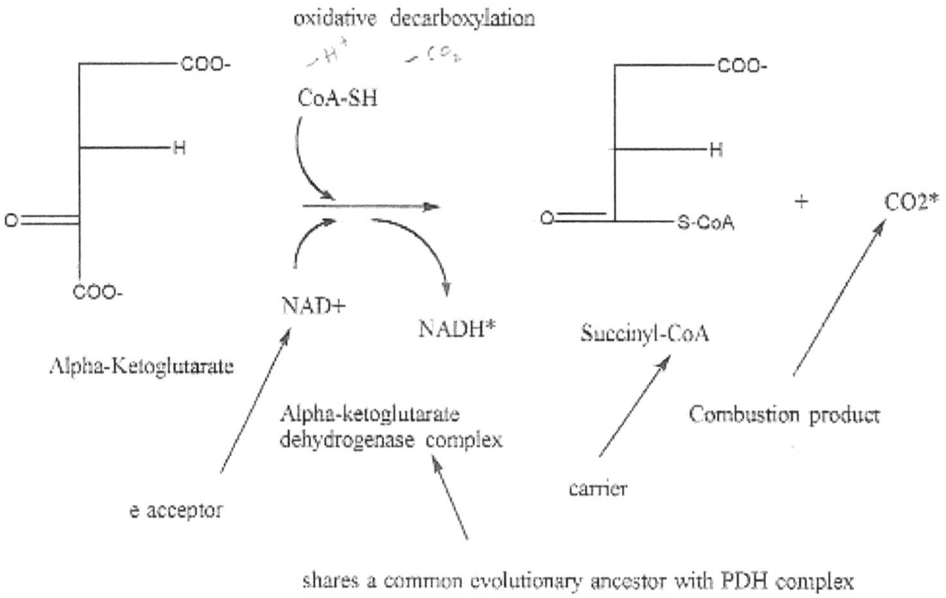

oxidative decarboxylation

CoA-SH

$-H^+$ $-CO_2$

COO-

H

O

COO-

Alpha-Ketoglutarate

NAD+

NADH*

Succinyl-CoA

COO-

H

S-CoA

+   CO2*

Combustion product

e acceptor

Alpha-ketoglutarate
dehydrogenase complex

carrier

shares a common evolutionary ancestor with PDH complex

For details see PDH reaction mechanism

Notes:

1

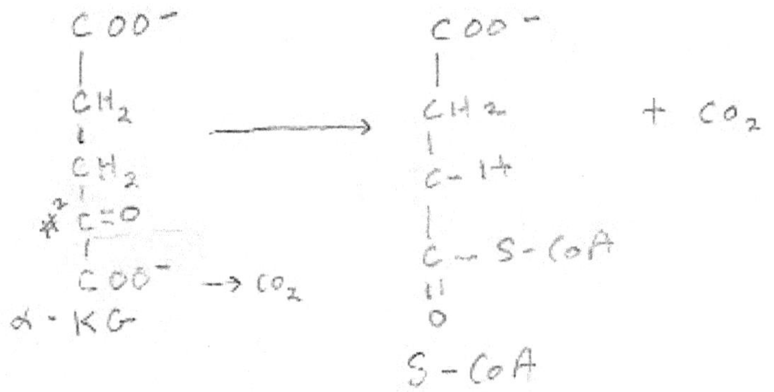

$$COO^- \quad\quad\quad COO^-$$
$$CH_2 \quad\quad\quad CH_2 \quad\quad + CO_2$$
$$CH_2 \quad\longrightarrow\quad C-H$$
$$*^2 C=O \quad\quad\quad C-S-CoA$$
$$COO^- \rightarrow CO_2 \quad\quad O$$
$$\alpha-KG \quad\quad\quad S-CoA$$

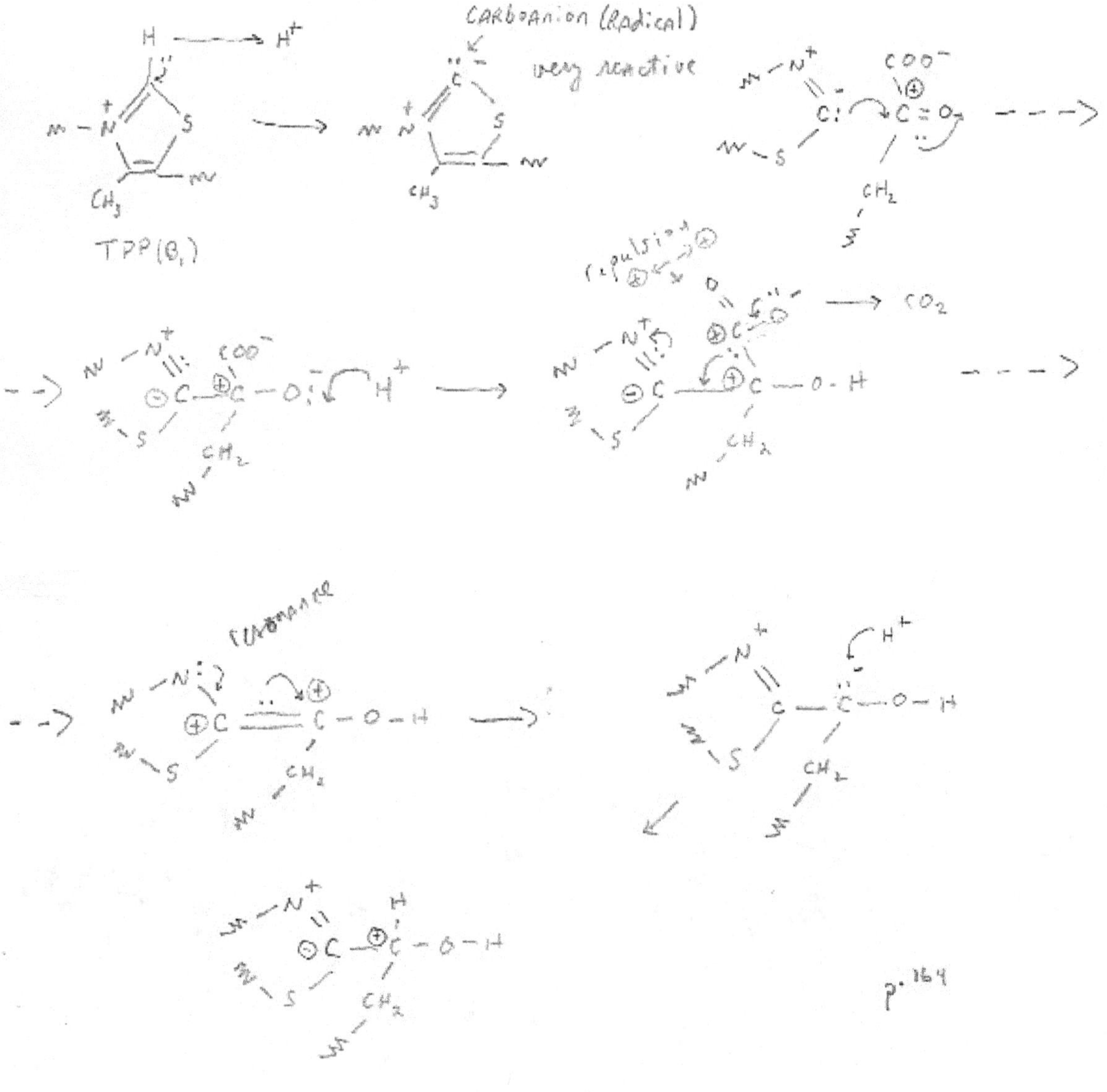

CARBOANION (Radical)

very reactive

TPP (B₁)

repulsion

resonance

p. 164

2

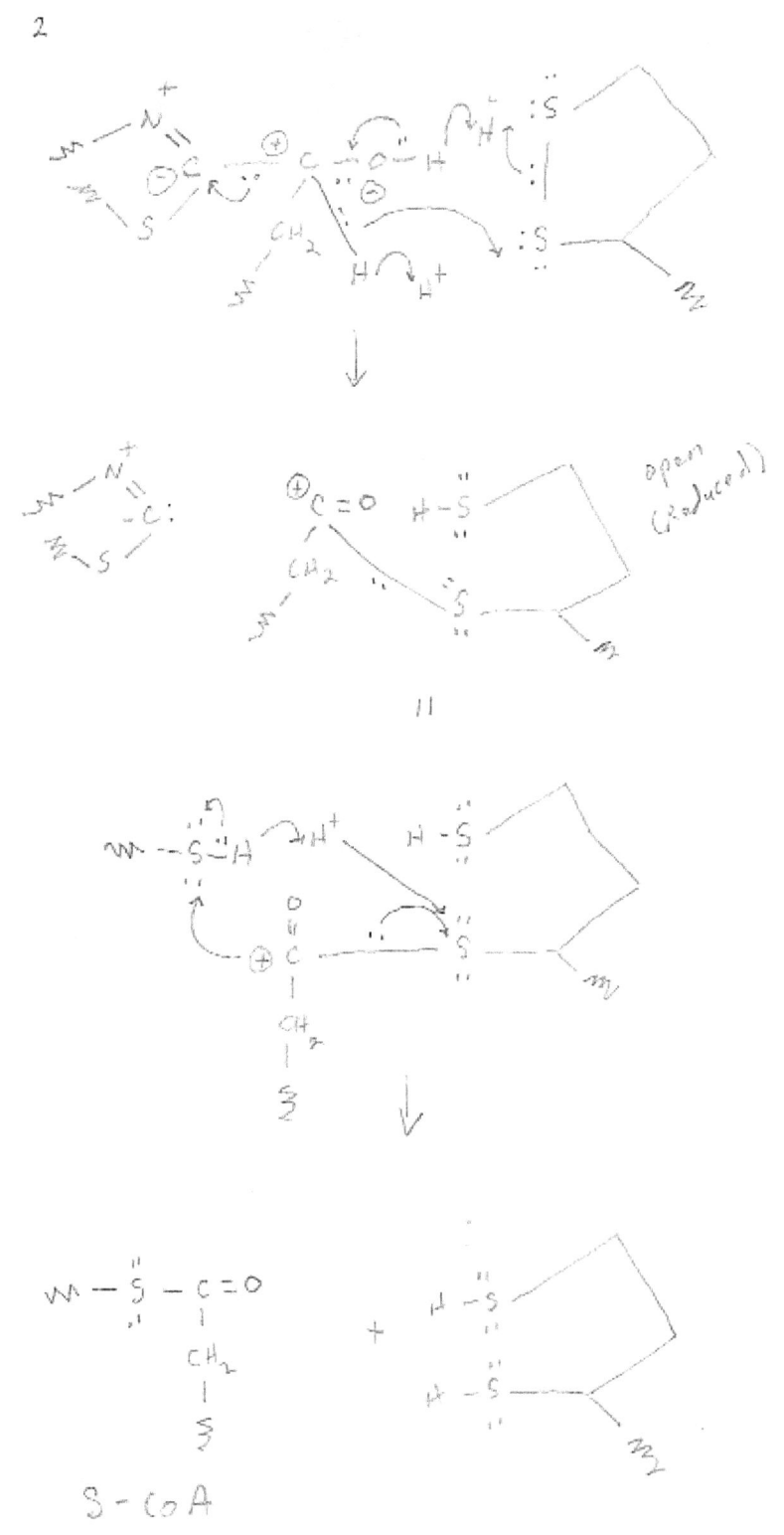

open
(reduced)

$\parallel$

S-CoA

3

NAD+   NADH $+ H^+$

Vitamin B1 de-protonates to carboanion which is very reactive.
This reacts with the keto function of a-KG: a carbonyl addition.
The carbonyl function radicalizes to add a proton to become an alcohol.
Through resonance , this releases CO_2 and stabilizes the molecule: decarboxylation
The carbon of alcohol becomes a carboanion and protonates.
This same carbon de-protonates twice in the presence of sulfur atoms.
CoA-SH deprotonates and reduces the sulfur atom and covalently bonds to carbonyl function.
Thus creating Succinyl-CoA.
The sulfur atoms oxidize to close it's ring structure liberating neutral H atom.
This neutral H atom can divide into a p+ and e- .
The NAD+ can absorb this division which is equal to 1 p+ (C:-)and 2 e- (N+) : oxidative
The p+ balance is absorbed by the hydroxyl ion of water producing water.

Review pages 168 Lysine and 171-2 NAD+.

This CO_2 and H_2O production is biology's way of combustion. Very different than CO_2 and H_2O produced by fire: total destruction!

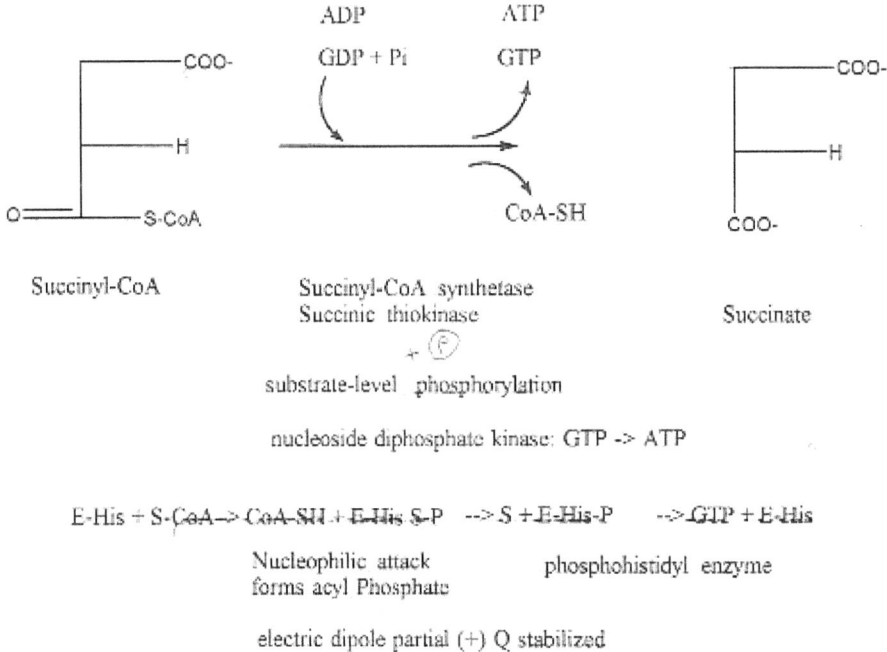

Succinyl-CoA   Succinyl-CoA synthetase   Succinate
                  Succinic thiokinase

substrate-level phosphorylation

nucleoside diphosphate kinase: GTP -> ATP

E-His + S-CoA -> CoA-SH + E-His S-P   --> S + E-His-P   --> GTP + E-His

           Nucleophilic attack   phosphohistidyl enzyme
           forms acyl Phosphate

           electric dipole partial (+) Q stabilized

Notes:

1.

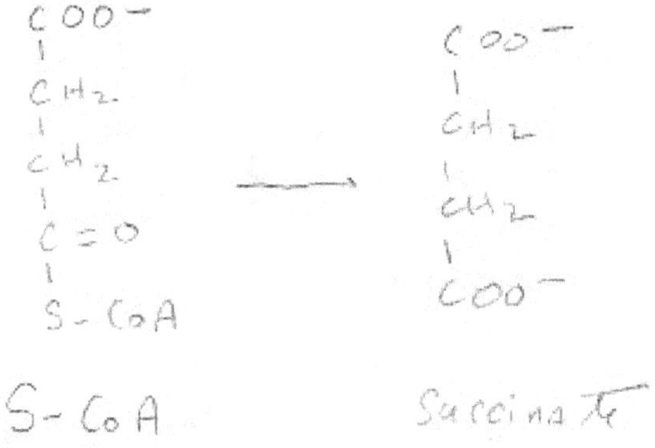

COO^-
|
CH_2
|
CH_2
|
$C = O$
|
$S - CoA$

$\longrightarrow$

COO^-
|
CH_2
|
CH_2
|
COO^-

S-CoA   Succinate

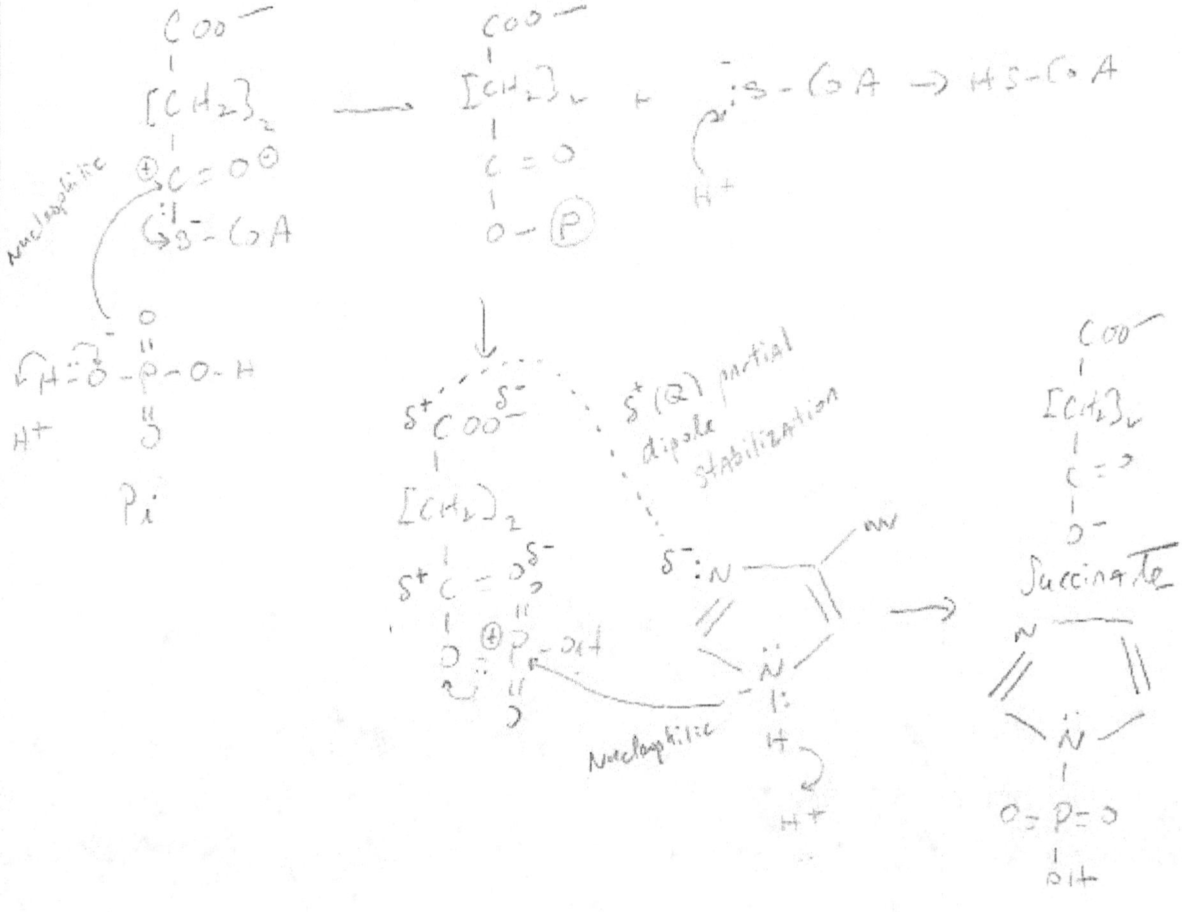

COO^-
|
$[CH_2]_2$
|
$\oplus C = O \ominus$
|
$\ominus S - CoA$ Nucleophilic

$\longrightarrow$

COO^-
|
$[CH_2]_v$
|
$C = O$
|
$O - \textcircled{P}$

$+$ $\bar{:}S - CoA \rightarrow HS - CoA$ H^+

$H:\ddot{\ddot{O}} - \overset{O}{\underset{O}{\overset{||}{P}}} - O - H$ H^+

Pi

$\delta^+ \overset{\delta^-}{COO^-}$
|
$[CH_2]_2$
|
$\delta^+ C = O^{\delta^-}$
|
$O \oplus P - OH$ Nucleophilic

$\delta^{+(\oplus)}$ partial dipole stabilization

$\delta^- : N$

H^+

H^+

$\longrightarrow$ Succinate

COO^-
|
$[CH_2]_v$
|
$C = O$
|
O^-

$O = P = O$
|
OH

2.

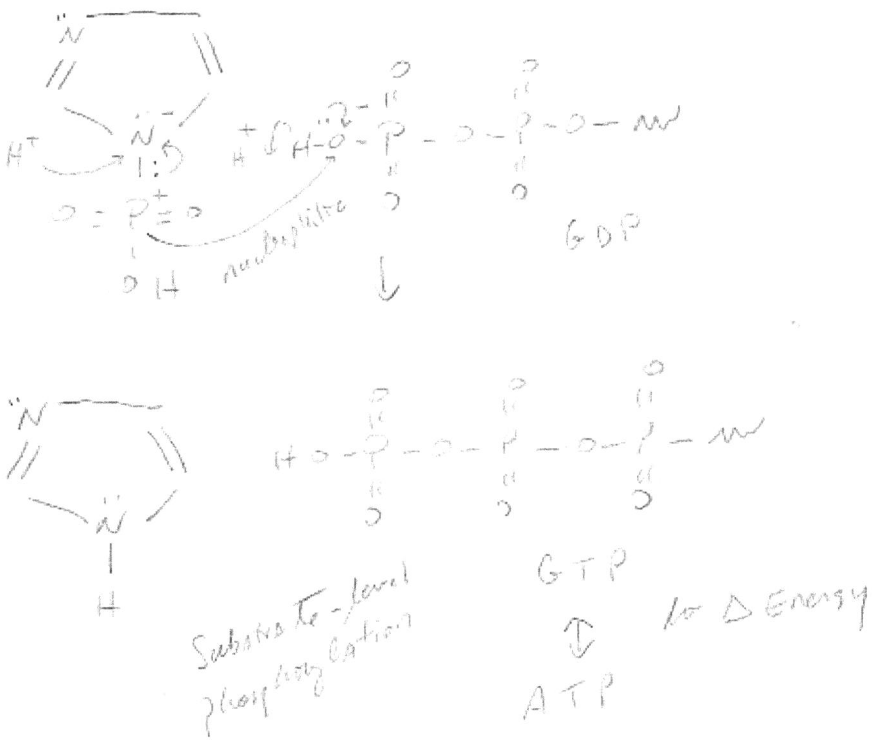

Substrate-level phosphorylation

GDP

GTP

$\updownarrow$

ATP

to Δ Energy

Notes:

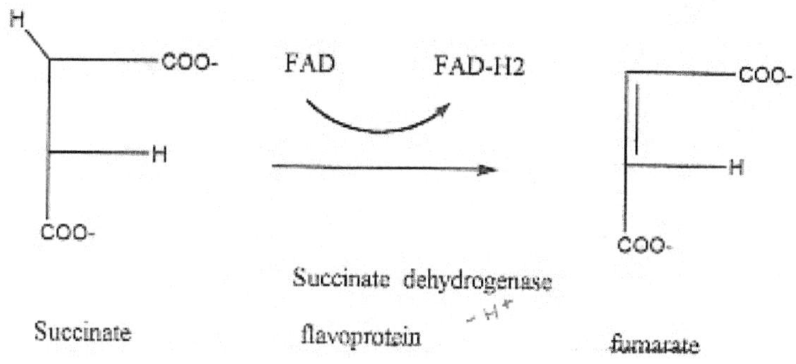
COO-
H
COO-
FAD
FAD-H2
COO-
H
COO-
Succinate
Succinate dehydrogenase
flavoprotein
- H+
fumarate

inner mitochondrial membrane

e -> O2 : 1.5 ATP molecules / e par

Notes:

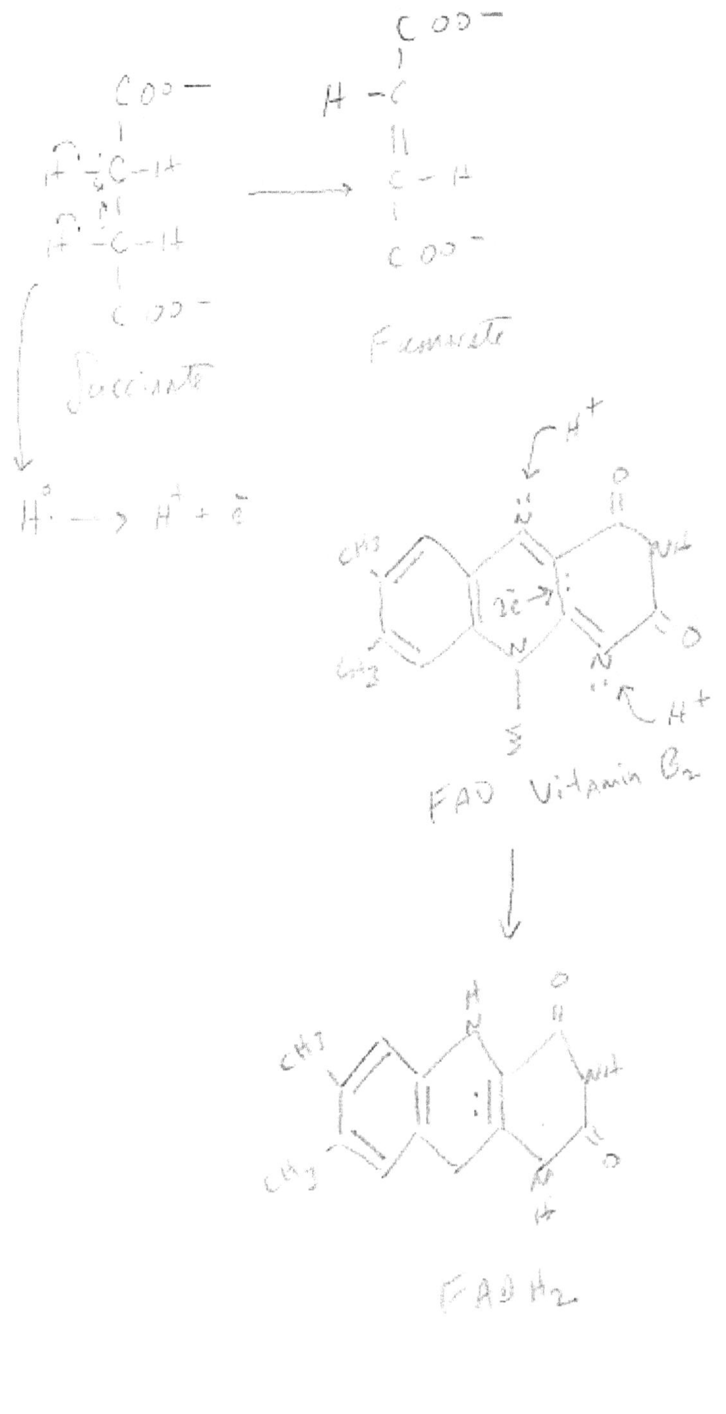

Succinate

Fumarate

FAD Vitamin B₂

FADH₂

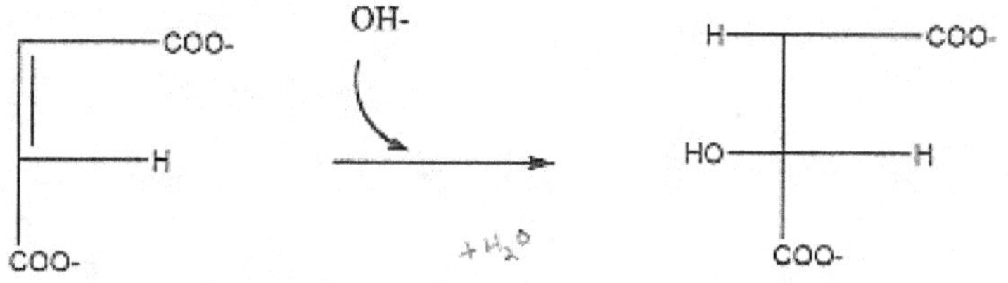

OH-

+H₂O

Fumarase

fumarate

Malate

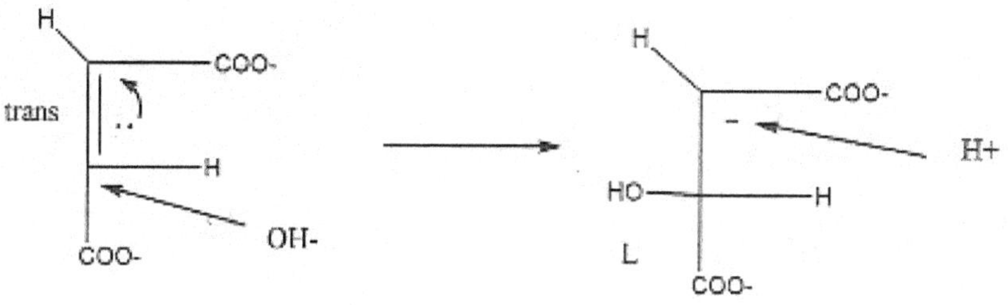

trans

trans DB only = stereospecific

In reverse, L-Malate is substrate

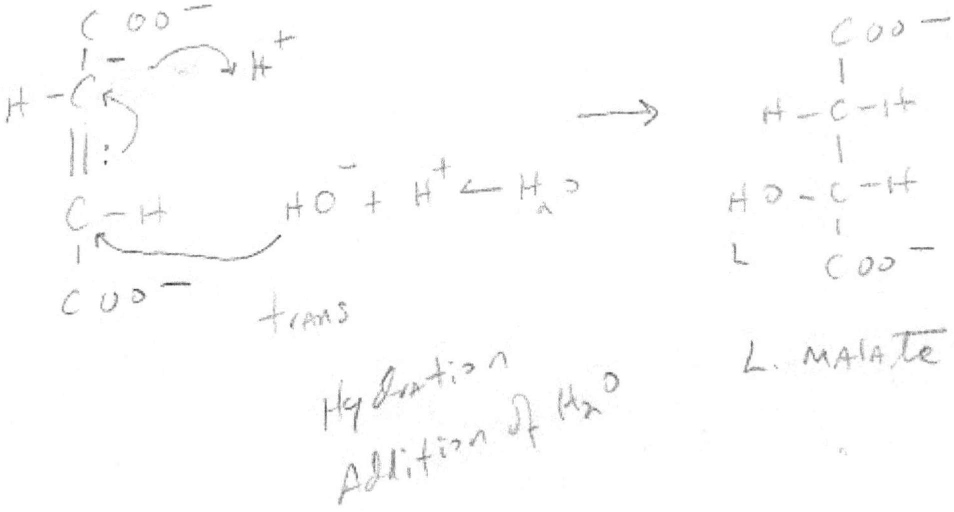

H+

Carbanion transition state

trans

Hydration
Addition of H₂O

L. MALATE

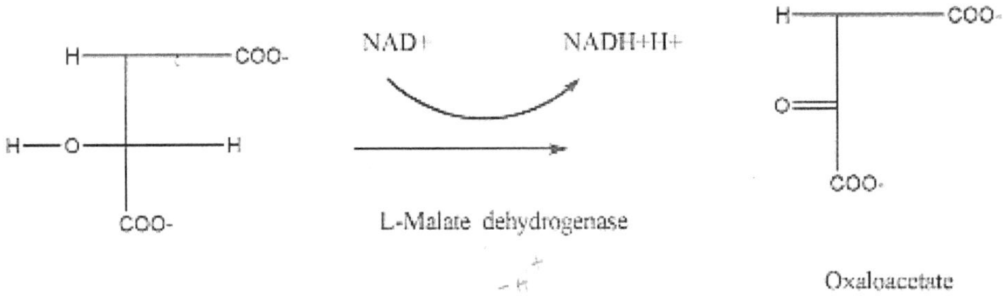

Malate

L-Malate dehydrogenase

Oxaloacetate

Repeat cycle

Cycle removes 2C as CO2 and generates Reduced compounds: NADH + FADH2 which carry e to the oxidative phosphorylation pathway with the final e acceptor O2 of water.

Compound combusted to $CO_2 + H_2O$ + Energy (ATP) + Heat (Infrared radiation) attenuated by the specific heat of water.

Notes:

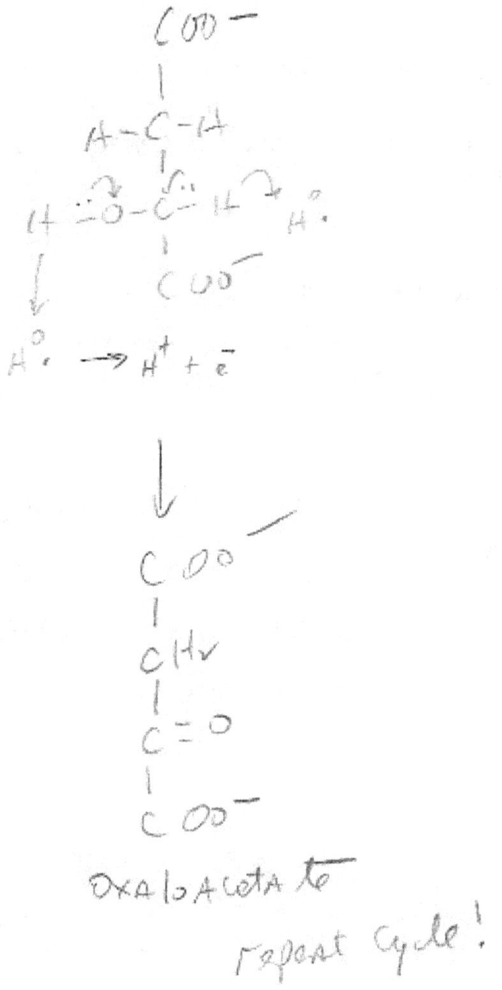

$$COO^-$$
$$|$$
$$A-C-H$$
$$|$$
$$H-O-C-H \quad H^{\bullet}$$
$$|$$
$$\downarrow \quad COO^-$$
$$H^{\bullet}_O \longrightarrow H^+ + \bar{e}$$

$$\downarrow$$

$$C\;OO$$
$$|$$
$$CH_2$$
$$|$$
$$C=O$$
$$|$$
$$COO^-$$

OXALOACETAte

repeat cycle!

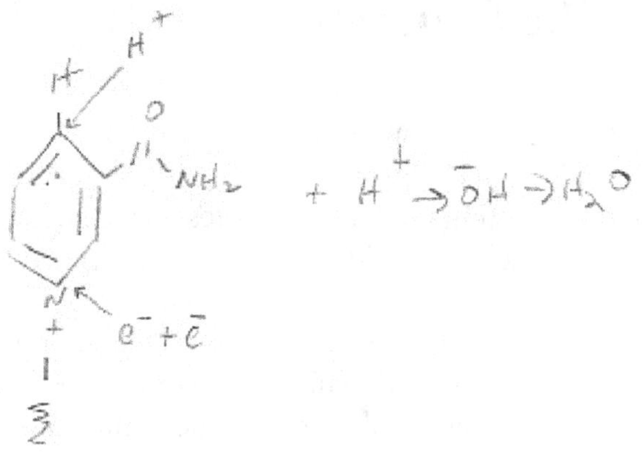

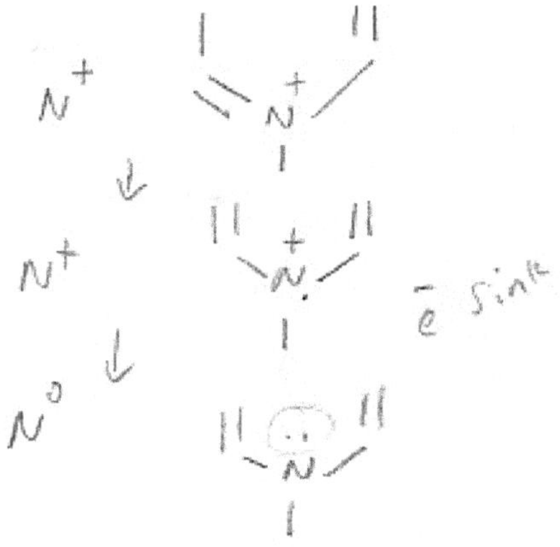

$$+ H^+ \rightarrow \bar{O}H \rightarrow H_2O$$

$$e^- + \bar{e}$$

$$N^+$$

$$N^+$$

$$N^0$$

$$\bar{e} \; sink$$

Chapter 11

The Generation of H2O: The Second ½ of Combustion

Oxidative phosphorylation is the process of reducing O2 with H. The reducing H is donated by NADH and FADH2. Remember: H(0)=p+ + e- H:- = p+ + 2e- H+ = p+. The process, biological oxidation-reduction reactions, creates a transmembrane H+ potential that drives ATP synthesis. ATP supplies the energy for biological processes (energy transduction).

1. The Q e- is done by membrane bound carriers.
2. Q e- → downhill = exergonic → free energy coupled to uphill transport of H+ → transmembrane electrochemical potential.
3. Q H+ is down gradient → ATP (Chemiosmotic Theory).

e- → e- acceptors

C-H reduced	+ NAD+ oxidized	→	C oxidized	+ NADH + H+ reduced	Catabolic
"	+ NADP+		"	NADP+ + H+	Anabolic
	FAD			FADH or FADH2	
	FMN			FMNH FMNH2	

B5 = Niacin NAD
B1 = Riboflavin FAD

Membrane Bound Carriers: Integral proteins with prosthetic groups.
3 types of e- transfers:

1. Direct : F3+ → Fe2+.
2. H(0) = H+ + e-
3. H:- = H+ + 2e-

Reducing equivalent = 1 e-

Respirator chain: NAD/FAD + Q + cytochromes (Fe-proteins + Hemes) + Fe-S proteins + O2

NAD/FAD → Q → Fe-S Proteins → Cytochromes → O2.

Ubiquinone = coenzyme Q = Q

$:\ddot{O}: \rightarrow :\ddot{II}: \ \overset{\frown}{e} \rightarrow :\ddot{I}: \quad :\ddot{O}: +e^-$

resonance $\rightarrow$

$CH_3O \quad \overset{5}{\underset{}{\bigcirc}} \quad (CH_2-CH=\overset{CH_3}{\underset{|}{C}}-CH_2)_{\overline{10}} H$

$CH_3O \quad \quad CH_3$

oxidized

$:\ddot{II}: \rightarrow :\ddot{II}:$
$:\ddot{O}: \quad :\ddot{O}:^- CH^+$

radical
oxonium ion
uncoupled ē pair

$:\ddot{II}:? \quad \overset{\frown}{H^+}$
$:\ddot{O}: \leftarrow$

$\frac{1}{2} H^+ + e^-\cdot = H°$ (splits)

$:\ddot{O}:\overset{\frown}{-} H^+ + \bar{e}$

Semiquinone radical
$\cdot QH$
Semi-reduced

$:\ddot{O}\overset{\frown}{-} H$

$\frac{1}{2} H^+ + \bar{e} \cdot = H°$

$O:\overset{\frown}{-}H$

Ubiquinol
QH_2
reduced

$O:\overset{\cdot\cdot}{-}H$

Cytochromes
prosthetic group = porphyrin

HEME   conjugation Absorbs visible
                        light

A,B,C Differ create Types:

Fe protoporphyrin IX   b-type
Heme C   c-type
Heme A   a-type

Directs:

$$N-Fe^{+3} \overset{e^-}{\longrightarrow} N-Fe^{+2}-N$$

oxidized   Reduced

At the center is an Fe core. Fe partakes in a direct Redox reaction which is stable. Of interest, the Heme is found in Hemoglobin, the Red Blood Cell. It gives the RBC it's color. But more importantly, it carries O2 and CO2.

$$\overset{3d}{\qquad} \qquad \overset{4s}{\qquad}$$

Fe [Ar] ↑↓ ↑↑↑↑   ↑↓   [Ar] $3d^6 4s^2$

18
Ar
$3s^2 3p^6$

d^{10}:

$d^{1\,2\,3\,4\,5\,6\,7\,8\,9\,10\,11}$

$1s^2 2s^2 2p^6 3s^2 3p^6 3d^6 4s^2$

36
Kr
$4s^2 4p^6$

$$\overset{3d}{\qquad}$$

Fe [Ar] ↑↓ ↑↑↑↑ Fe^{+2}

Fe [Ar] ↑ ↑↑↑↑ Fe^{+3} $+ \ \bar{e}$

Direction of oxidation # ↑↓ $+ \longleftrightarrow (-)$

O# ↑ = oxidation

O# ↓ = Reduction

$(-) \longleftrightarrow O \longleftrightarrow (+)$

reduction   oxidation

Fe^{+2} Fe^{+3}

gain $\bar{e}$ loss $\bar{e}$

Electronic Configuration shows 3d atomic structure. Clarification of the direction of the Redox reaction involving Fe.

Notes:

Fe-S Centers

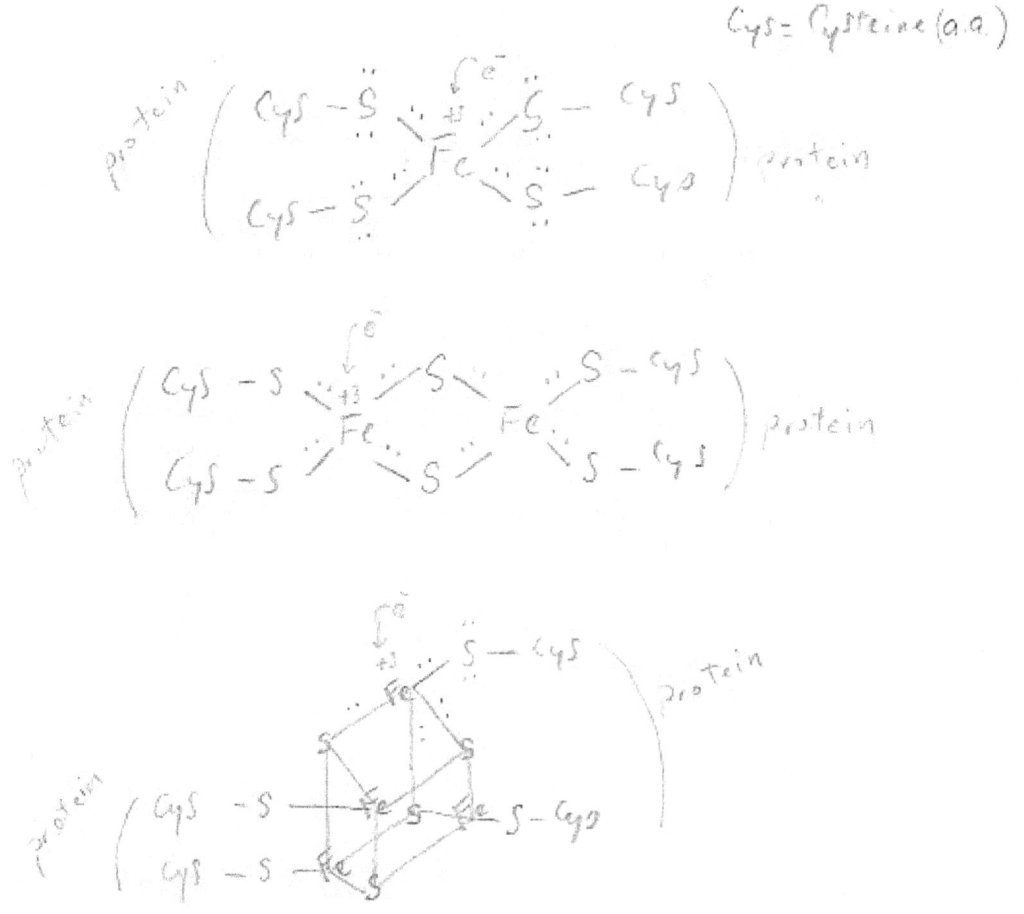
Cys = Cysteine (a.a.)
protein
Cys
S
Fe
+3

The Heme and Fe-S centers are throughout the oxidation phosphorylation chain.
At the core is the Fe+3 → Fe+2 Oxidation-Reduction reaction.

reduction   *oxidation*

By Accepting an ē, you oxidize the Another.
    lost ē

gain ē   accepting an ē

Spontaneous
lower → higher

TABLE 19-2   Standard Reduction Potentials of Respiratory Chain and Related Electron Carriers

Reduction half

Redox reaction (half reaction)	E'° (V)
$2H^+ + 2e^- \longrightarrow H_2$	−0.414
$NAD^+ + H^+ + 2e^- \longrightarrow NADH$	−0.320
$NADP^+ + H^+ + 2e^- \longrightarrow NADPH$	−0.324
NADH dehydrogenase (FMN) + $2H^+$ + $2e^- \longrightarrow$ NADH dehydrogenase (FMNH$_2$)	−0.30
Ubiquinone + $2H^+$ + $2e^- \longrightarrow$ ubiquinol	0.045
Cytochrome b (Fe^{3+}) + $e^- \longrightarrow$ cytochrome b (Fe^{2+})	0.077
Cytochrome c_1 (Fe^{3+}) + $e^- \longrightarrow$ cytochrome c_1 (Fe^{2+})	0.22
Cytochrome c (Fe^{3+}) + $e^- \longrightarrow$ cytochrome c (Fe^{2+})	0.254
Cytochrome a (Fe^{3+}) + $e^- \longrightarrow$ cytochrome a (Fe^{2+})	0.29
Cytochrome a_3 (Fe^{3+}) + $e^- \longrightarrow$ cytochrome a_3 (Fe^{2+})	0.35
$\frac{1}{2}O_2 + 2H^+ + 2e^- \longrightarrow H_2O$	0.8166

final   best reducing

$Fe^{+3} = +.771 > Fe^{+2} = -.440$

Better to accept   less
an ē

oxidized   reduced

oxidizing
easily reduced

Oxidation + reduction are interchangeable terms. It all depends on ē flow.

Oxidation: giving up electrons.
Reduction: accepting electrons.
A reducing agent oxidizes.
An oxidizing agent reduces.

Oxidation ½ reaction: NADH + H+ → NAD+ + 2H+ + 2e   NADH gives up 2 e. 0 → + O# up
Reduction ½ reaction: Ubiquinone + 2H+ + 2e → Ubiquinol Ubiquinone accepts 2 e. 0 → (-) O# down.
Over-all reaction Oxidation-Reduction: NADH + H+ + Ubiquinone → NAD+ + Qubiquinol

---

1   Cox, Michael, Nelson, David, **Lehninger Principles of Biochemistry,** 4th Edition, p. 695.

2

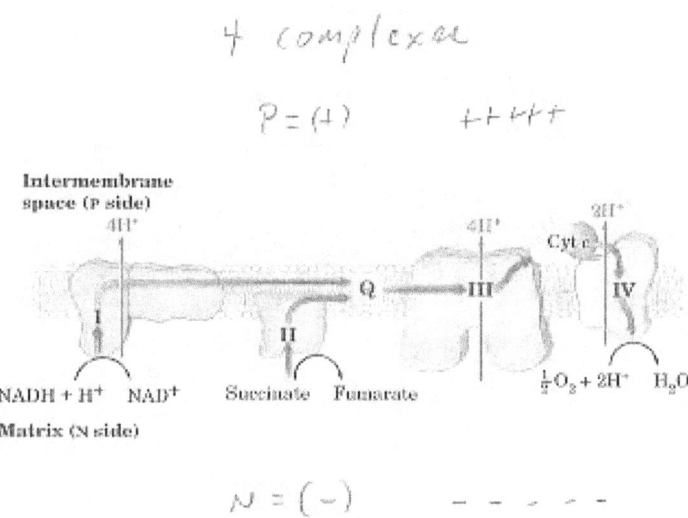

Semi-permeable membrane separates the Matrix and Intermembrane space.

10 H+ are pumped into the Intermembrane space making this area more positive compared to the Matrix.

The 4 complexes are in series according to their reduction potential.

Complex I : NADH Dehydrogenase or NADH:Ubiquinone Oxidoreductase

NADH → H:- → FMN → Fe-S → N2 (Fe-S) → Q → QH2 + 4H+
      Hydride   X2   2e per 4H+

Electrochemical potential conserves energy from e transfer reaction.
This energy conservation drives ATP synthesis.
ATP provides potential energy for all energy transductions throughout the organism.

Q → QH2 = 2e; 2X2= 4 therefore QH2 goes twice and produces 4 H+.

How Niacin and Riboflavin handles the e has been reviewed.
It seems that the transfer process is at the 2 e rate. But knowing that B2 and Q can have a semi-state begs the question of a 1 or 2 e rate.

2   Cox, Michael, Nelson, David, **Lehninger Principles of Biochemistry,** 4th Edition, p. 703.

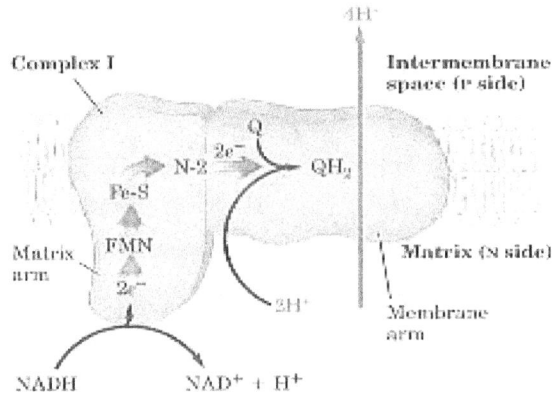

Complex II : Succinate Dehydrogenase (Kreb's cycle)

Succinate binding site → FAD → 3 2Fe-2S centers → Q   a 2e transfer.

Heme b acts as a stray e sink to prevent O2 radical formation: ROS (Reactive Oxygen Species): H2O2 (Hydrogen Peroxide) and Superoxide Radical: O2(-) 1e.

Succinate, FAD, and Fe-S are outside of the membrane.

3   Cox, Michael, Nelson, David, **Lehninger Principles of Biochemistry,** 4th Edition, p. 698.

4

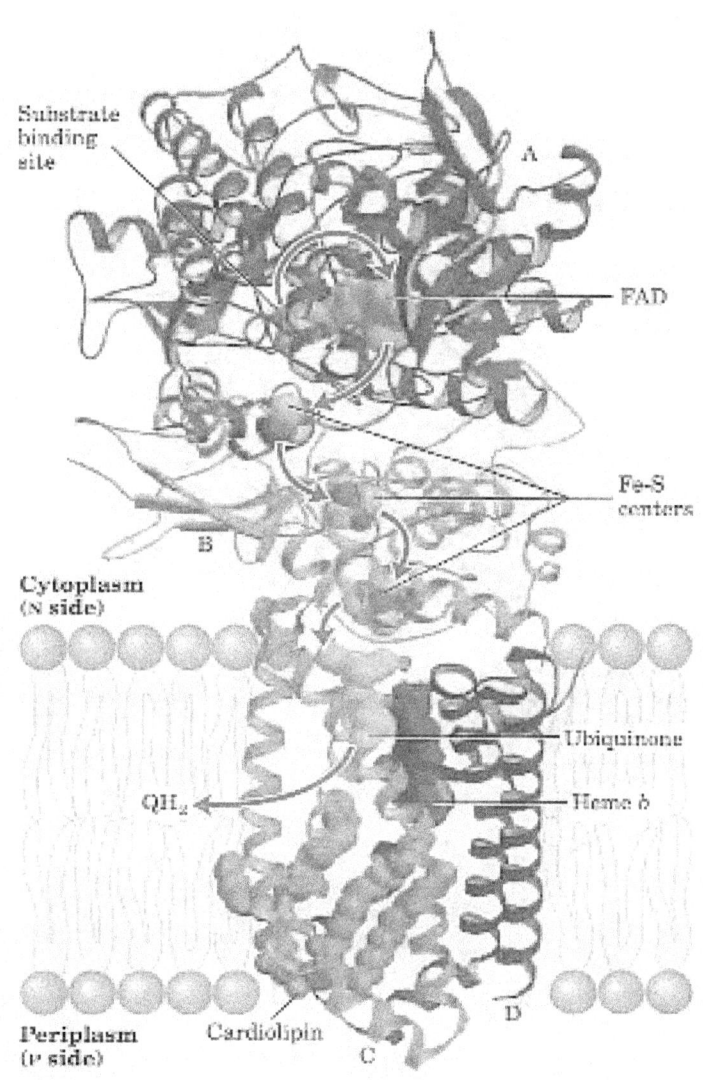

Substrate
binding
site

FAD

Fe-S
centers

Cytoplasm
(N side)

B

Ubiquinone

QH$_2$

Heme b

Periplasm
(p side)

Cardiolipin

C

D

A

4   Cox, Michael, Nelson, David, **Lehninger Principles of Biochemistry,** 4th Edition, p. 699.

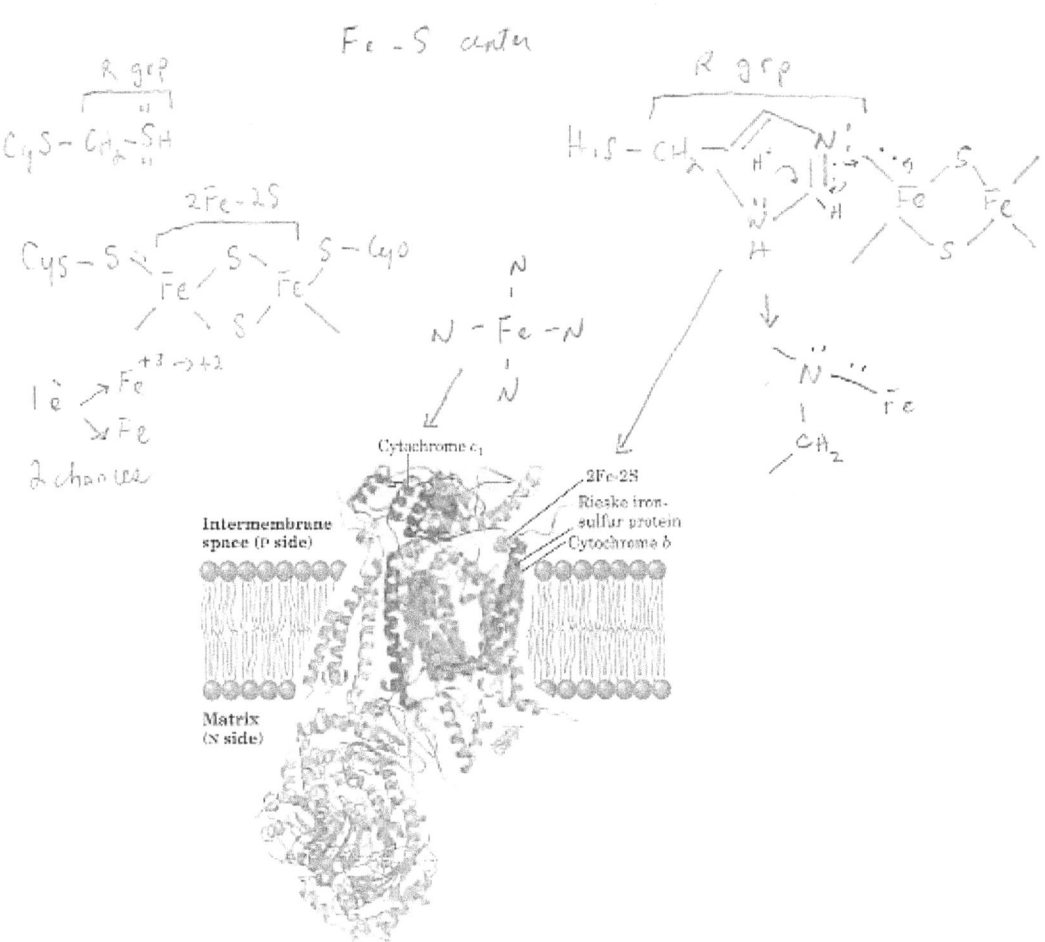

The Rieske iron-sulfur protein attaches by the amino acid, Histadine.

5   Cox, Michael, Nelson, David, **Lehninger Principles of Biochemistry,** 4th Edition, p. 700.

Complex III : Cytochrome bc1 complex or ubigquinone:cytochrome c oxidoreductase

6

Q(P) and Q(N) are 2 binding sites for Q, which can be in 3 different states: QH2, Q, and Q.-.
Approximate locations of the important sites.

6   Cox, Michael, Nelson, David, **Lehninger Principles of Biochemistry**, 4th Edition, p. 700.

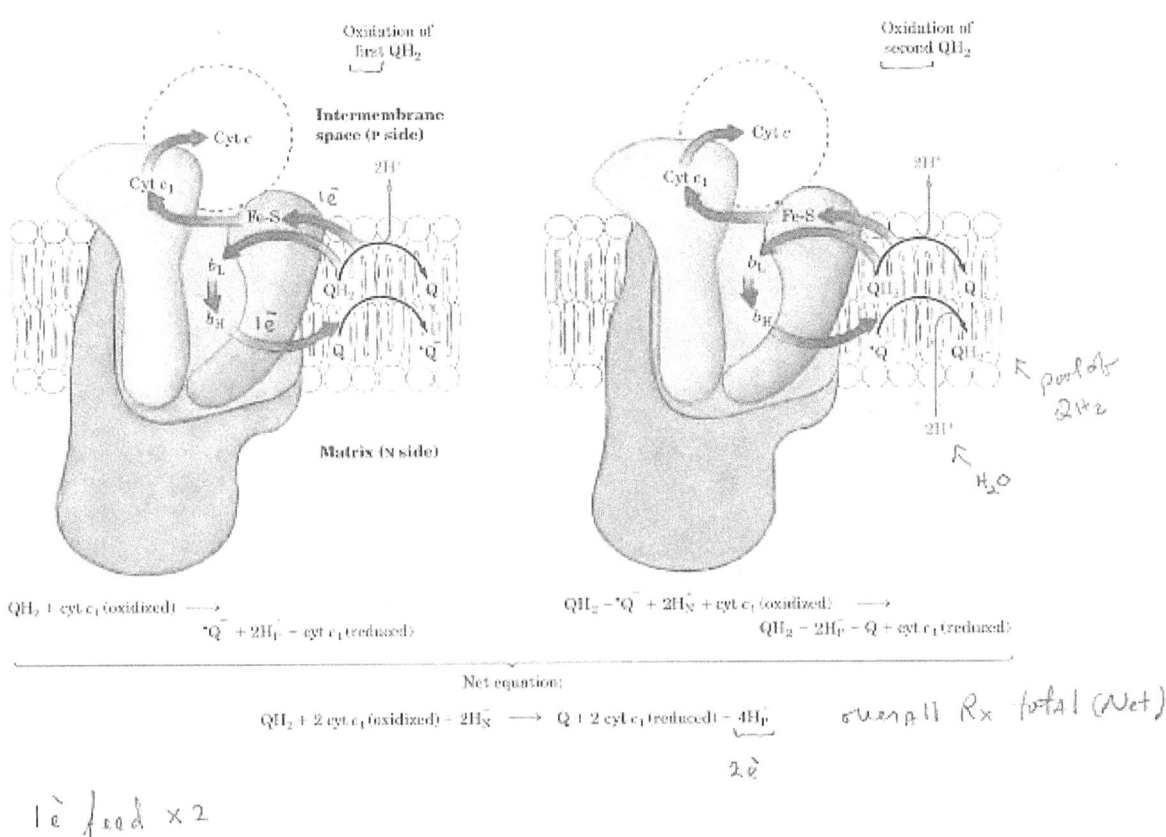

QH₂ + cyt c₁ (oxidized) ⟶
'Q⁻ + 2H⁺_P + cyt c₁ (reduced)

QH₂ - 'Q⁻ + 2H⁺_N + cyt c₁ (oxidized) ⟶
QH₂ - 2H⁺_P - Q + cyt c₁ (reduced)

pool of 2H₂

H₂O

Net equation:

QH₂ + 2 cyt c₁ (oxidized) + 2H⁺_N ⟶ Q + 2 cyt c₁ (reduced) + 4H⁺_P

overall Rx total (Net)

2e⁻

1e⁻ feed × 2

Complex IV : Cytochrome Oxidase

Cytochrome c → O2 → H2O

3 subunits: I : 2 heme + Cu (Fe-Cu center).
II : 2 Cu + Cys (-SH).
III : unknown?

7   Cox, Michael, Nelson, David, **Lehninger Principles of Biochemistry,** 4th Edition, p. 701.

8
9

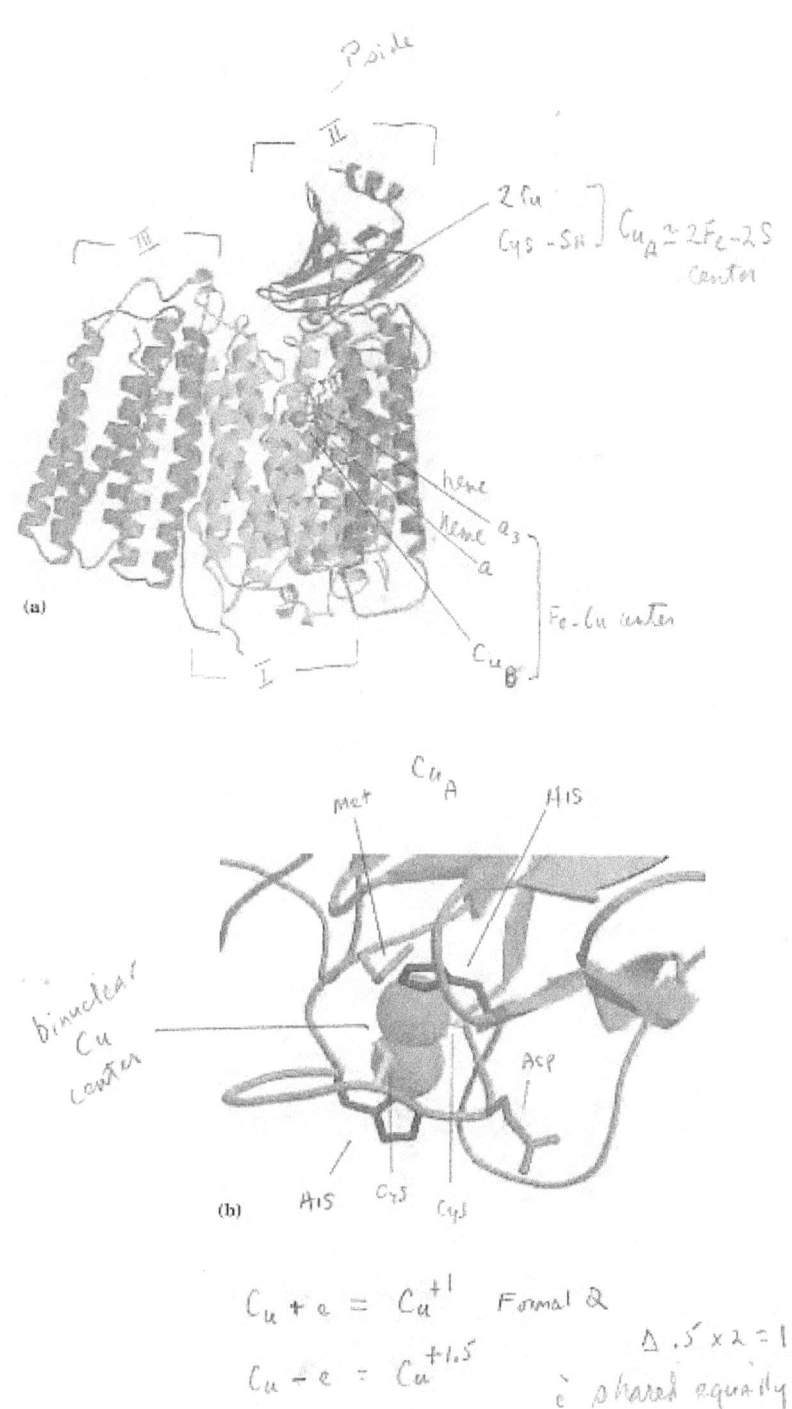

$$Cu + e = Cu^{+1} \quad \text{Formal 2}$$
$$Cu - e = Cu^{+1.5}$$

$\Delta .5 \times 2 = 1$
$\bar{e}$ shared equally

8   Cox, Michael, Nelson, David, **Lehninger Principles of Biochemistry,** 4th Edition, p. 702.
9   Cox, Michael, Nelson, David, **Lehninger Principles of Biochemistry,** 4th Edition, p. 702.

10

1e at a time!

gradient (unknown?)

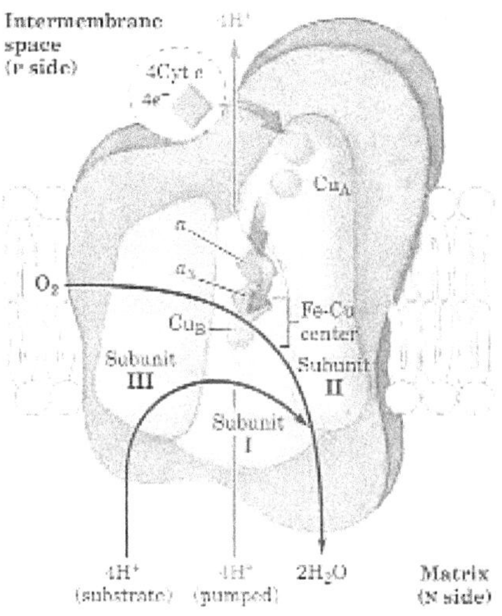

$2H_2 = 4H^+$
$4e^-$

$O_2^{-2} = peroxy$ derivative

$O_2^{-2} + 2e \rightarrow O_2^{-4} + 4H^+ \rightarrow 2H_2O$

10 Cox, Michael, Nelson, David, **Lehninger Principles of Biochemistry,** 4th Edition, p. 702.

11

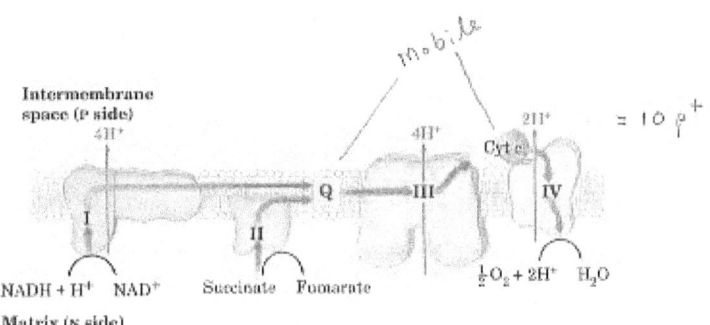

12

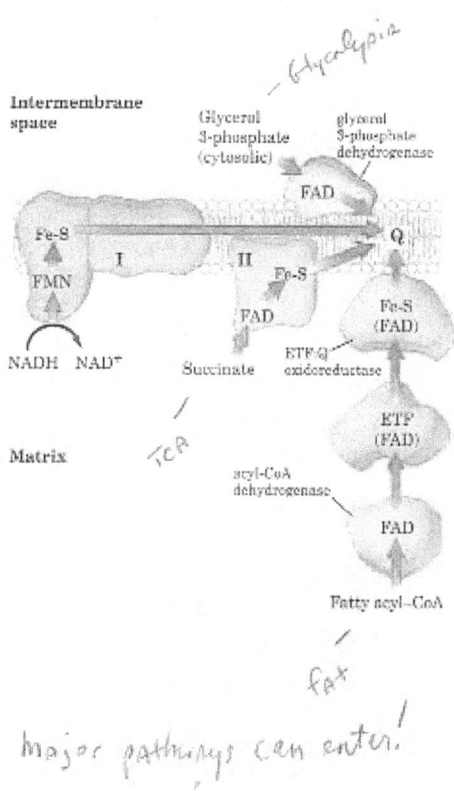

major pathways can enter!

11 Cox, Michael, Nelson, David, **Lehninger Principles of Biochemistry,** 4th Edition, p. 703.
12 Cox, Michael, Nelson, David, **Lehninger Principles of Biochemistry,** 4th Edition, p. 697.

ATP Synthesis:

The energy from the transfer of e is conserved in and creates a electrochemical gradient. The force inherent in this gradient is called proton-motive force. This energy or force will be the source for the creation of ATP, the energy currency of the cell. ATP is available for any form of cellular work to be done.

[13]

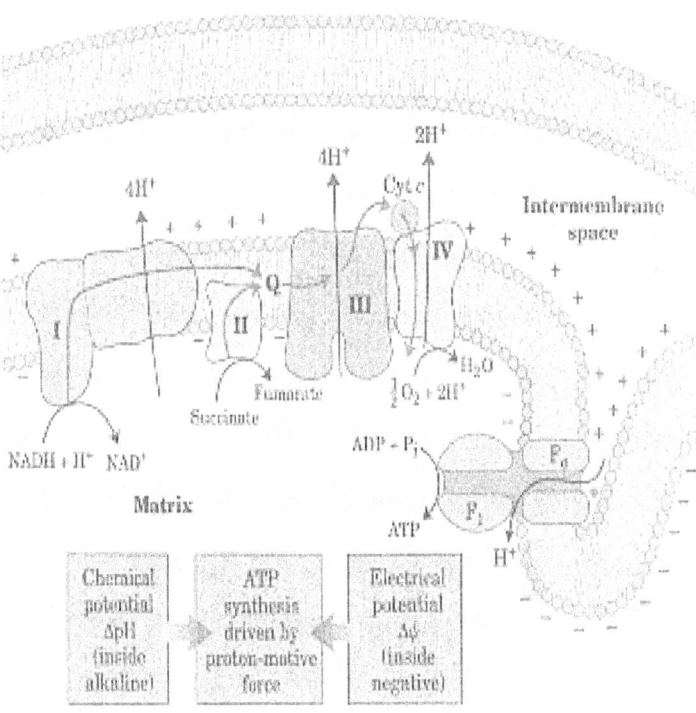

Chemiosmosis: H+ → delta pH + e- → delta psi
F0 = p+ transport channel
F1 = ATP synthesis enzyme.

13 Cox, Michael, Nelson, David, **Lehninger Principles of Biochemistry,** 4th Edition, p. 705.

2 Pathways for Only 1 Goal: C and e Pathways for the Synthesis of ATP

The 3 Main Stages

Stage 1 : **Acetyl-CoA Production**

Acetyl-CoA is an activated 2C molecule. All 3 monomer nutrients, amino acids, fatty acids, and glucose, can convert to this molecule.
During the process, electrons are produced.
In Glycolysis, the catabolism of glucose, Pyruvate is reached. Pyruvate converts to Acetyl-CoA with the release of CO_2 and e. Pyruvate is a 3C molecule. CO_2 is the fully oxidized C.

a.a. (digested protein) $\rightarrow$ A-CoA + e

f.a. (digested fat) $\rightarrow$ A-CoA + e

Glu (digested carbohydrate) $\rightarrow$ Pyruvate (3C) $\rightarrow$ A-CoA (2C) + CO_2 + e

So A-CoA is the 2C molecule of choice.

Stage II : **Acetyl-CoA Oxidation**

TCA cycle conserves matter and energy because of it's cyclic behavior. A-CoA (2C) adds to Oxaloacetate (4C) to produce Citrate (6C) molecule.
From Citrate to OA, 2 CO_2 and e are produced.

Citrate $\rightarrow$ OA + 2 CO_2 + e

All e are carried by e carriers which are reduced in the process.

e + NAD+/FADH $\rightarrow$ NADH + H+/FADH2
H:- = Hydride ion

Stage III : **e Transfer and Oxidative Phosphorylation**

e enter and get transferred in the e-transfer chain. The e end up in water. ATP is synthesised.

$2H+ + \frac{1}{2} O_2 \rightarrow H_2O$ $1/2O_2 = O$
ADP + Pi $\rightarrow$ ATP* $\rightarrow$ cell work (Thinking, running, breathing, etc)

At the carbo-loading session, the football player consumes a carbohydrate-enriched meal. The above scheme is one of catabolism. The 3 macronutrients are digested to monomers and then catabolized to CO_2 and e. The e are used to create ATP. ATP is the energy available for all types of biological work. In football, this is mainly thinking, running, and breathing. The session is designed to fore-fill the energy needed to play a game for football. We found that our players tolerated the meal well, no GI complaints. They were alert and enduring. The details of the program was discussed earlier.

1

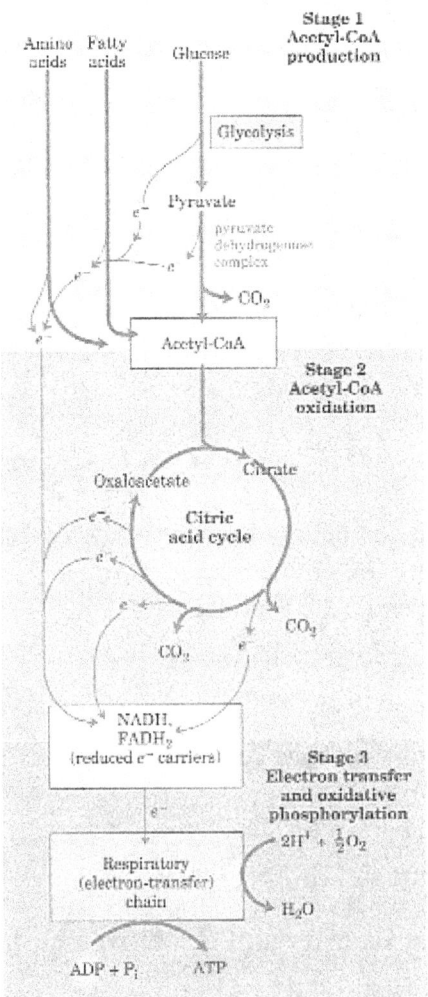

---

1 Cox, Michael, Nelson, David, **Lehninger Principles of Biochemistry**, 4th Edition, p. 602.

Glycolysis

Glucose oxidation

There are 2 phases : 1. Preparatory
2. Payoff

Preparatory Phase

Glucose (6C) uses 2 ATPs to create 2 (3C) structures. The phosphate moieties are high energy compounds.

Payoff Phase

Oxidation produces 2 NADH + H+
Phosphorylation produces 2 ATP. This is called substrate-level phosphorylation. This is repeated once creating 2 more ATP.

Net: 2 NADH + H+ $\rightarrow$ Oxidative Phosphorylation under aerobic (oxygenated) condition.
4 - 2 = 2 ATP

Glucose(6C) $\rightarrow$ 2 (3C)Pyruvate

Notes:

2

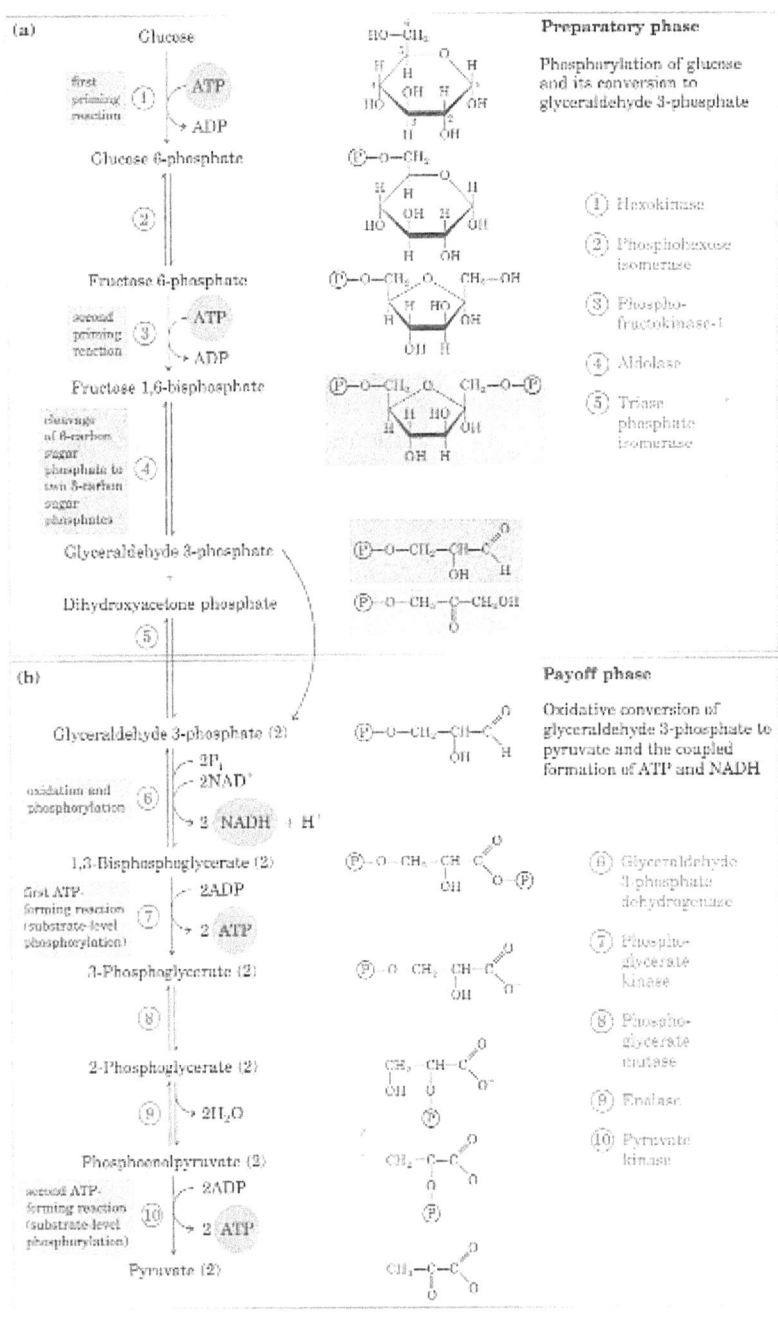

2   Cox, Michael, Nelson, David, **Lehninger Principles of Biochemistry,** 4th Edition, p. 524.

Kreb's Cycle: TCA

Acetyl-CoA Oxidation

Oxaloacetate (4C) joins Acetyl-CoA (2C) to produce Citrate (6C). Through 2 Oxidative Decarboxylation reactions, 2 CO_2 and 2 NADH are formed.

Succinyl-CoA (4C) undergoes a Substrate-level Phosphorylation to create 1 GTP = ATP.

A Dehydrogenation reaction produces 1 $FADH_2$.

Eventually, OA is re-created (cyclic).

Everything is doubled because 1 turn of the cycle represents only ½ of glucose = 1 Acetyl-CoA.

OA + A-CoA $\rightarrow$ C

C (2OdeCO2) $\rightarrow$ 2 NADH + 2 CO_2

S-CoA (1S-LP) $\rightarrow$ 1 GTP + S

S (1 DeH) $\rightarrow$ $FADH_2$ + F

M (1 DeH) $\rightarrow$ OA + 1 NADH

1 cycle: original substrate is re-created.

OdeCO2 :1) Oxidative : removes H with it's e.
2) Decarboxylation : removes CO_2.

S-LP : removes the PO4(-2) group and conserves the energy in ATP.

DeH : removes the H with it's e.

e(s) do not exist by themselves. They are associated with an atom. H atom is the one of choice in a given reaction.

3

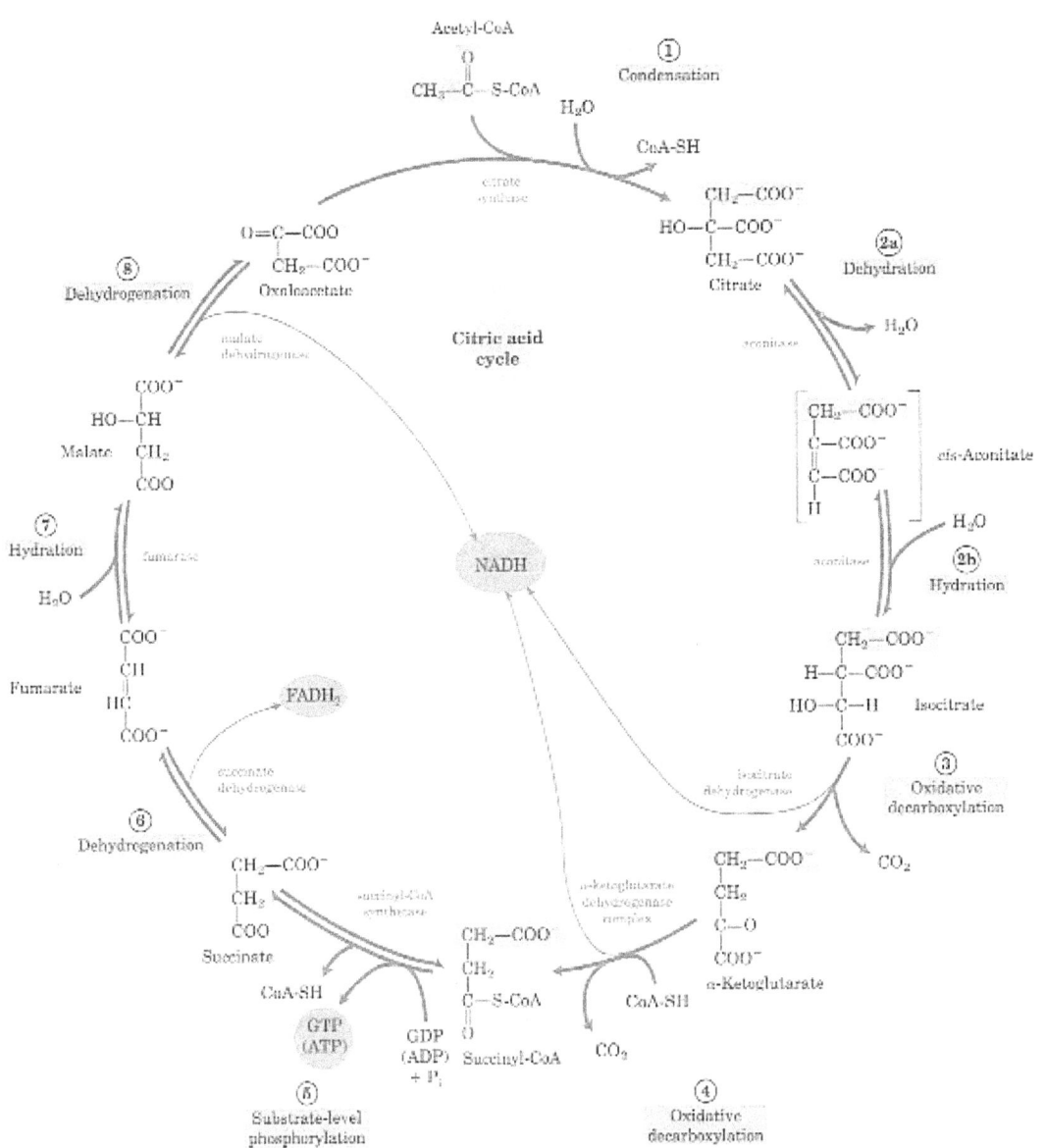

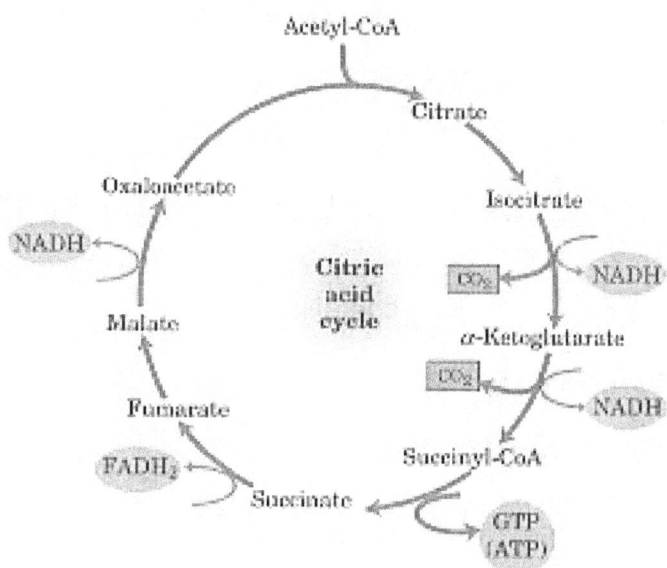

Summary : 1 turn of Kreb's cycle

3 NADH
1 FADH2
1 GTP

1 Glucose produces 2 turns of the Kreb's cycle.

TABLE 19–5   ATP Yield from Complete Oxidation of Glucose

Process	Direct product	Final ATP
Glycolysis	2 NADH (cytosolic)	3 or 5*
	2 ATP	2
Pyruvate oxidation (two per glucose)	2 NADH (mitochondrial matrix)	5
Acetyl-CoA oxidation in citric acid cycle (two per glucose)	6 NADH (mitochondrial matrix)	15
	2 FADH$_2$	3
	2 ATP or 2 GTP	2
Total yield per glucose		30 or 32

*The number depends on which shuttle system transfers reducing equivalents into the mitochondrion.

4   Cox, Michael, Nelson, David, **Lehninger Principles of Biochemistry,** 4th Edition, p. 615.
5   Cox, Michael, Nelson, David, **Lehninger Principles of Biochemistry,** 4th Edition, p. 716.

Note: 1 NADH → 2.5 ATP
1 FADH2 → 1.5 ATP

Also the conversion of Pyr → A-CoA yields 2 NADH
During Glycolysis, ATP production depends on the reducing equivalents used to shuttle e into the mitochondria.

Beta-Oxidation

Fat Oxidation

In Beta-Oxidation, a long chain f.a. is oxidized 2C at a time.

f.a. → nA-CoA + FADH2 + NADH+H+

This process is cyclic.

A-CoA → Kreb's cycle.

FADH2 + NADH+H+ → O.P.

All resulting in ATP production on a much larger scale.

Beta means the 2nd C from the carbonyl function of the activated f.a..

If the f.a. is 16 C atoms long, this will yield 8 (2C) A-CoA.

8 (2C)A-CoA → 16 CO2 and 64 e in the Kreb's cycle.

These e proceed through O.P.

The e from Beta-Oxidation carried by FADH2 and NADH+H+ enter O.P. to yield ATP.

6   Cox, Michael, Nelson, David, **Lehninger Principles of Biochemistry,** 4th Edition, p. 637.

7

(a)

(C_{18}) $R-CH_2-\overset{\beta}{CH_2}-\overset{\alpha}{CH_2}-\overset{}{C}-S\text{-}CoA$
$\qquad\qquad\qquad\qquad\qquad\overset{\|}{O}$ Palmitoyl CoA

acyl CoA dehydrogenase $\left\{\begin{array}{l}\text{- FAD}\\ \rightarrow \text{FADH}_2\end{array}\right.$

$\quad\quad\quad\quad\underset{\text{H}}{\overset{\text{H}}{|}}$
$R-CH_2-C=C-C-S\text{-}CoA$
$\quad\quad\quad\quad\underset{\text{H}}{|}\ \underset{\text{O}}{\|}$ $trans\text{-}\Delta^2\text{-}$
$\qquad\qquad\qquad\qquad\qquad$ Enoyl CoA

enoyl-CoA hydratase $\left\{\text{- H}_2\text{O}\right.$

$\quad\quad\quad\overset{\text{OH}}{|}$
$R-CH_2-C-CH_2-C-S\text{-}CoA$
$\quad\quad\quad\underset{\text{H}}{|}\qquad\quad\underset{\text{O}}{\|}$ L β Hydroxy-
$\qquad\qquad\qquad\qquad\qquad$ acyl-CoA

β-hydroxyacyl-CoA dehydrogenase $\left\{\begin{array}{l}\text{- NAD}^+\\ \rightarrow \text{NADH + H}^+\end{array}\right.$

$R-CH_2-C-CH_2-C-S\text{-}CoA$
$\qquad\quad\underset{\text{O}}{\|}\qquad\quad\underset{\text{O}}{\|}$ β-Ketoacyl-CoA

acyl CoA acetyltransferase (thiolase) $\left\{\text{- CoA-SH}\right.$

(C_{14}) $R-CH_2-C-S\ CoA\ +\ CH_3-C-S\ CoA$
$\qquad\qquad\quad\underset{\text{O}}{\|}\qquad\qquad\qquad\underset{\text{O}}{\|}$

(C_{14}) Acyl-CoA $\qquad$ Acetyl -CoA
(myristoyl-CoA)

(b)

C_{14} ◯ ⟶ Acetyl -CoA
C_{12} ◯ ⟶ Acetyl -CoA
C_{10} ◯ ⟶ Acetyl -CoA
C_8 ◯ ⟶ Acetyl -CoA
C_6 ◯ ⟶ Acetyl -CoA
C_4 ◯ ⟶ Acetyl -CoA

Acetyl -CoA

---

7   Cox, Michael, Nelson, David, **Lehninger Principles of Biochemistry,** 4th Edition, p. 638.

TABLE 17-1   Yield of ATP during Oxidation of One Molecule of Palmitoyl-CoA to CO_2 and H_2O

Enzyme catalyzing the oxidation step	Number of NADH or $FADH_2$ formed	Number of ATP ultimately formed*
Acyl-CoA dehydrogenase	7 $FADH_2$	10.5
β-Hydroxyacyl-CoA dehydrogenase	7 NADH	17.5
Isocitrate dehydrogenase	8 NADH	20
α-Ketoglutarate dehydrogenase	8 NADH	20
Succinyl-CoA synthetase		8†
Succinate dehydrogenase	8 $FADH_2$	12
Malate dehydrogenase	8 NADH	20
Total		108

*These calculations assume that mitochondrial oxidative phosphorylation produces 1.5 ATP per $FADH_2$ oxidized and 2.5 ATP per NADH oxidized.

†GTP produced directly in this step yields ATP in the reaction catalyzed by nucleoside diphosphate kinase (p. XXX).

Almost 4X the ATP production from 1 f.a. compared to 1 Glucose.

Amino Acid Oxidation

In the text, I talked about gluconeogenesis in detail. This pathway is energy costly. The oxidation of a.a. takes a more energy-saving route via the Kreb's cycle.

8   Cox, Michael, Nelson, David, **Lehninger Principles of Biochemistry,** 4th Edition, p. 640.

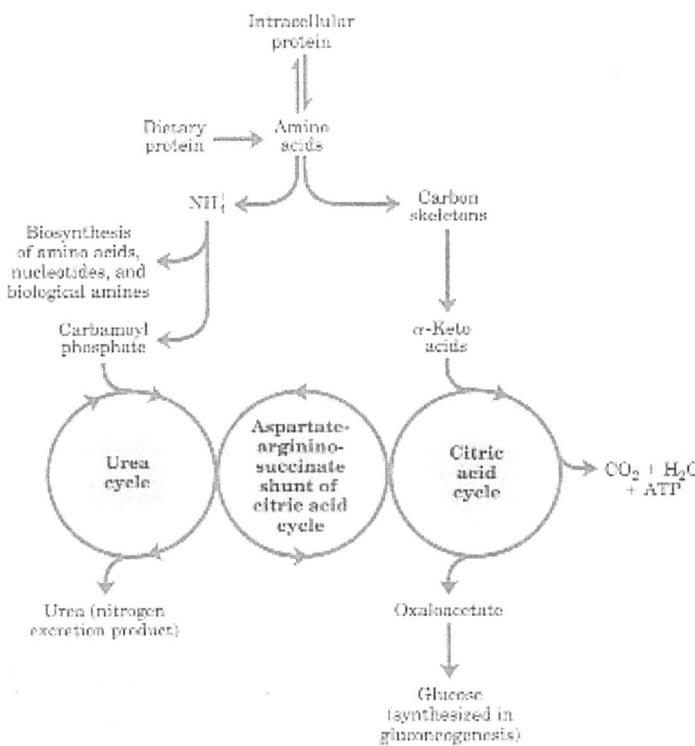

This is an overview of the destiny of an a.a.. The right side of the diagram is the important side in this discussion.

The a.a. is converted to a carbon skeleton intermediate, alpha-keto acid, which can enter the Kreb's cycle directly. This leads to the complete combustion of the a.a. which yields CO2 + H2O + ATP.

The a.a. can be converted to O.A. via Pyr. and then Glucose, for synthetic purposes or if need be, for catabolic purposes, such as in starvation.

---

9   Cox, Michael, Nelson, David, **Lehninger Principles of Biochemistry,** 4th Edition, p. 657.

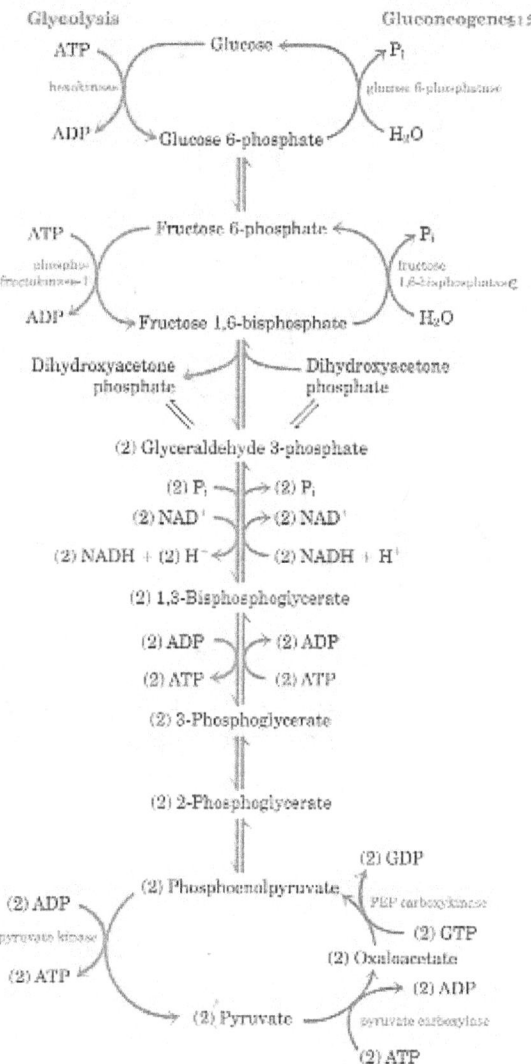

Energy is paid in the lost of ATP, NADH+H+, and Pi generation. Many of these reverse steps in Glycolysis requires energy to proceed.

10 Cox, Michael, Nelson, David, **Lehninger Principles of Biochemistry,** 4th Edition, p. 544.

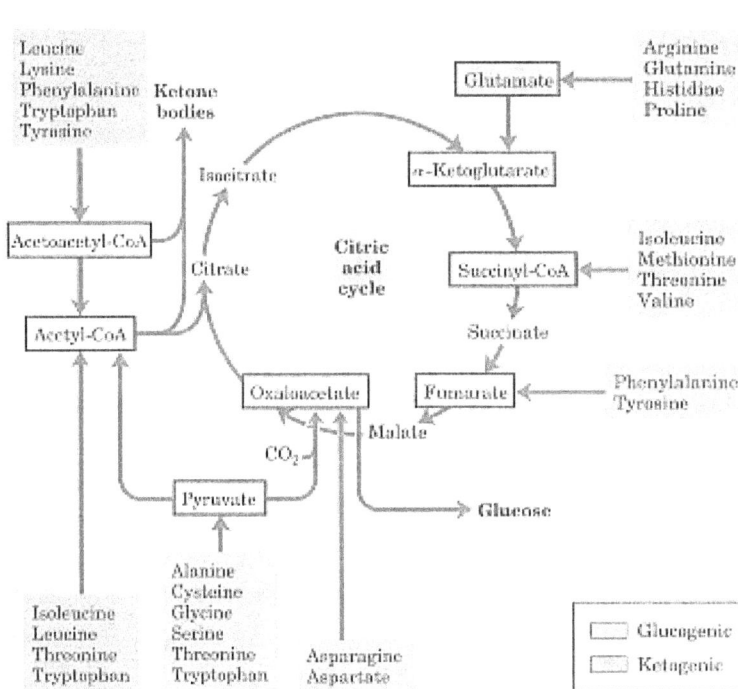

Amino acids can generate glucose or ketones. Most are glucogenic. In this book, we discussed the fate of Alanine. In the above scheme, entering directly into the Kreb's cycle, Alanine adds to CO2 to generate O.A.. This compound joins A-CoA in the Kreb's cycle.

11 Cox, Michael, Nelson, David, **Lehninger Principles of Biochemistry,** 4th Edition, p. 671.

Chapter 13

Vitamins, Ginseng, and Royal Jelly

Multiple Vitamins, Minerals, and Trace Metals

Introduction

During the hydration session, we provided a multiple vitamin and Ginseng with Royal Jelly.
The vitamins provided the co-enzymes for carbohydrate catabolism.
The Ginseng and Royal Jelly provided mental alertness during fatigue states.
This section will present an overview and background on these compounds.
The B vitamins have been discussed in the text. The B's are considered stress vitamins because of their activity in carbohydrate metabolism.
Much of the evidence for Ginseng and Royal Jelly are anecdotal in nature. The Chinese have used them for over a thousand years. I was mainly interested in the delivery of oxygen to the brain during the fatigue of the 4th quarter of football. A single tablet lasted approximately 4 hours.
The players tolerated the tables and described alertness, endurance, and energy well-being during the entire game. One player stated that he "could play another game!"

Vitamins

Lipid soluble vitamins: KEAD
Water soluble vitamins: B

A : Eyes
C: Antioxidant
D: bones and Teeth ($Ca2+$)
E: Antioxidant
K: Coagulation
B1 Thiamin: (CHOH Metabolism)
B2 Riboflavin: ($FADH_2$)
B5 Niacin : ($NADH + H+$)
Folic Acid: Blood
B12: Blood, Nervous System

Centrum® Adults Under 50

Supplement Facts

Serving Size 1 Tablet

Each Tablet Contains	% Daily Value
Vitamin A 3,500 IU (29% as Beta-Carotene)	70%
Vitamin C 60 mg	100%
Vitamin D 400 IU	100%
Vitamin E 30 IU	100%
Vitamin K 25 mcg	31%
Thiamin 1.5 mg	100%
Riboflavin 1.7 mg	100%
Niacin 20 mg	100%
Vitamin B_6 2 mg	100%
Folic Acid 400 mcg	100%
Vitamin B_{12} 6 mcg	100%
Biotin 30 mcg	10%
Pantothenic Acid 10 mg	100%
Calcium 200 mg	20%
Iron 18 mg	100%
Phosphorus 20 mg	2%
Iodine 150 mcg	100%

Biotin: ? (Kreb's cycle, f.a.)
B6: Pantothenic Acid : ? Growth, Skin

Minerals

Calcium: Teeth, Bones
Iron: RBC
Phosphorus: Renal, Metabolism
Iodine: Thyroid
Magnesium: Metabolism
Zinc: Growth, Hair, Skin, GI, Cornea
Selenium: (Liver Metabolism)
Copper: RBC
Manganese: Liver and Muscle
Chromium: (CHOH Metabolism)
Molybdenum: Growth
Chloride: Renal, Metabolism
Potassium: Renal, Electrical

Trace Metals

Anecdotal
Many Claims
Very little science

Boron: ?
Nickel: ?
Silicon: ?
Tin: ?
Vanadium: ?

The B vitamins are used for Carbohydrate Metabolism.
The antioxidants are used to eliminate radical production.

Notes:

2

Magnesium 50 mg	13%
Zinc 11 mg	73%
Selenium 55 mcg	79%
Copper 0.5 mg	25%
Manganese 2.3 mg	115%
Chromium 35 mcg	29%
Molybdenum 45 mcg	60%
Chloride 72 mg	2%
Potassium 80 mg	2%
Boron 75 mcg	*
Nickel 5 mcg	*
Silicon 2 mg	*
Tin 10 mcg	*
Vanadium 10 mcg	*
* Daily Value not established.	

Ingredients: Calcium Carbonate, Potassium Chloride, Dibasic Calcium Phosphate, Magnesium Oxide, Microcrystalline Cellulose, Ascorbic Acid (Vit. C), Ferrous Fumarate, Pregelatinized Corn Starch, dl-Alpha Tocopheryl Acetate (Vit. E). **Contains < 2% of:** Acacia, Beta-Carotene, BHT, Biotin, Boric Acid, Calcium Pantothenate, Calcium Stearate, Cholecalciferol (Vit. D_3), Chromium Picolinate, Citric Acid, Corn Starch, Crospovidone, Cupric Sulfate, Cyanocobalamin (Vit. B_{12}), FD&C Yellow No. 6 Aluminum Lake, Folic Acid, Gelatin, Hydrogenated Palm Oil, Hypromellose, Manganese Sulfate, Medium-Chain Triglycerides, Modified Food Starch, Niacinamide, Nickelous Sulfate, Phytonadione (Vit. K), Polyethylene Glycol, Polyvinyl Alcohol, Potassium Iodide, Pyridoxine Hydrochloride (Vit. B_6), Riboflavin (Vit. B_2), Silicon Dioxide, Sodium Ascorbate, Sodium Benzoate, Sodium Citrate, Sodium Metavanadate, Sodium Molybdate, Sodium Selenate, Sorbic Acid, Stannous Chloride, Sucrose, Talc, Thiamine Mononitrate (Vit. B_1), Titanium Dioxide, Tocopherols, Tribasic Calcium Phosphate, Vitamin A Acetate (Vit. A), Zinc Oxide. **May also contain < 2% of:** Ascorbyl Palmitate, Maltodextrin, Sodium Aluminosilicate, Sunflower Oil.

Suggested Use: Adults — Take one tablet daily with food. Not formulated for use in children. Do not exceed suggested use.

As with any supplement, if you are pregnant, nursing, or taking medication, consult your doctor before use.

WARNING: Accidental overdose of iron-containing products is a leading cause of fatal poisoning in children under 6. Keep this product out of reach of children. In case of accidental overdose, call a doctor or poison control center immediately.

IMPORTANT INFORMATION:
Long-term intake of high levels of vitamin A (excluding that sourced from beta-carotene) may increase the risk of osteoporosis in adults. Do not take this product if taking other vitamin A supplements.

Ginseng

3

Ginseng is an adaptogen. It relieves stress. The plant contains saponins, soap-like compounds, and the root has a steroid-like compound.

There are 2 main effects:

1. Cooling or relaxing : Yin ginseng (American or Siberian species).
2. Warming or stimulating effects : Yang (Korean and Chinese).

Our football players used Siberian, Korean, and Chinese type ginsengs. They experienced a more stimulating effect overall. When the body warms, blood vessels dilate delivering more oxygen and blood to the organs, such as brain and muscles. Thus countering the effects of mental fatigue.

Anecdotal effects:

mental stimulation and physical activity
improved work performance
prevents fatigue
improves memory
strengthen the heart and nervous system
improves mental and physical vitality
improves resistance to disease
stimulates the endocrine system

It is recommended for many conditions. Athletic performance is one of them.

---

3   Schweikhart, Melissa, Vanderbilt University, Nashville, **Tennessee Health Psychology,** 115A, December 10, 1996.

Saponins may have a suppressive effect on the anxiety and fear caused by psychological stress.

[4]

A study of triathletes showed a prevention by the drug in the loss of physical performance characteristic of end of the season tiredness.

[5]

The active component of ginseng is the ginsenosides. This compound is steroid-like, triterpene saponins. If they act like steroids, then their action takes hours, days or weeks to take effect. What and how is the immediate effect of these compounds?

Royal Jelly

A gelatinous secretion of the worker bee that is fed to the queen bee to nutritionally support her.

A few studies to read.

Royal Jelly Studies

Antioxidant Properties of Royal Jelly Associated with Larval Age and Time of Harvest.

Liu JR, Yang YC, Shi LS, Peng CC.

bocky@sunws.nfu.edu.tw.

J Agric Food Chem. 2008 Nov 13.

A bee study.

4 Schepdael, Van P. (1993), Effect of Ginseng G115 on the physical condition of triathletes, **Acta Therapeutica,** 19(4), 337-347.
5 Leung, Kar Wah, Wong, Alice Sze-Tsai, Pharmacology of Ginsenoside: A Literature Review, **Chinese Medicine,** 2010, 5:20.

Royal Jelly Studies: Auto Immune Disorders

Honeybee royal jelly inhibits autoimmunity in SLE-prone NZB x NZW F1 mice

Mannoor MK, Shimabukuro I, Tsukamotoa M, Watanabe H, Yamaguchi K, Sato Y.

Department of Social and Environmental Medicine, Faculty of Medicine, University of the Ryukyus, Okinawa, Japan. Kaiissar@med.u-ryukyu.ac.jp

A mouse study.

Roayl Jelly Studies: Breast Cancer

Effect of royal jelly on bisphenol A-induced proliferation of human breast cancer cells.

PMID: 17213647 [PubMed - indexed for MEDLINE]

Royal Jelly Studies: Cognitive Enhancement

Royal Jelly Facilitates Restoration of the Cognitive Ability in Trimethyltin-Intoxicated Mice

Hattori N, Ohta S, Sakamoto T, Mishima S, Furukawa S.

229

Laboratory of Molecular Biology, Gifu Pharmaceutical University, 5-6-1, Mitahora-higashi, Gifu 502-8585, Japan. furukawa@gifu-pu.ac.jp.

Royal Jelly Studies: Liver Protection

The effects of royal jelly on liver damage induced by paracetamol in mice

Kanbur M, Eraslan G, Beyaz L, Silici S, Liman BC, Altinordulu S, Atasever A.

Department of Pharmacology and Toxicology, Faculty of Veterinary Medicine, University of Erciyes, Kayseri, Turkey.

Royal Jelly Studies: Osteoporosis

Royal jelly prevents osteoporosis in rats: beneficial effects in ovariectomy model and in bone tissue culture model.

Hidaka S, Okamoto Y, Uchiyama S, Nakatsuma A, Hashimoto K, Ohnishi ST, Yamaguchi M.

PMID: 16951718 [PubMed - in process]

Royal Jelly Studies: Cholesterol

Royal jelly supplementation improves lipoprotein metabolism in humans.

Guo H, Saiga A, Sato M, Miyazawa I, Shibata M, Takahata Y, Morimatsu F

R&D Center, Nippon Meat Packers, Inc., Ibaraki, Japan.

h.guo@nipponham.co.jp

Experientia. 1995 Sep 29;51(9-10):927-35

Effect of royal jelly on serum lipids in experimental animals and humans with atherosclerosis.

Vittek J.

Department of Medicine, New York Medical College, Valhalla 10595, USA.

PMID: 7556573 [PubMed - indexed for MEDLINE]

Royal Jelly Studies: Inflammation

Royal jelly inhibits the production of proinflammatory cytokines by activated macrophages.

Kohno K, Okamoto I, Sano O, Arai N, Iwaki K, Ikeda M, Kurimoto M.Fujisaki Institute, Hayashibara Biochemical Laboratories, Inc., Okayama, Japan.

kohnok@hayashibara.co.jp

PMID: 14745176 [PubMed - indexed for MEDLINE]

Royal Jelly Studies: Wound Healing

Augmentation of wound healing by royal jelly (RJ) in streptozotocin-diabetic rats.

Department of Pharmacology, Nihon University School of Dentistry, Matsudo, Japan.

PMID: 2391765 [PubMed - indexed for
MEDLINE]

Honey and royal jelly, like human milk, abrogate lectin-dependent infection-preceding Pseudomonas aeruginosa adhesion.

Lerrer B, Zinger-Yosovich KD, Avrahami B, Gilboa-Garber N.The Mina & Everard Goodman Faculty of Life Sciences, Bar-Ilan University, Ramat-Gan, Israel.

PMID: 18043624 [PubMed - indexed for MEDLINE]

Royal Jelly Studies - Diabetes

Royal jelly reduces the serum glucose levels in healthy subjects

Münstedt K, Bargello M, Hauenschild A.

Department of Obstetrics and Gynecology, Justus-Liebig University Hospital Giessen and Marburg, Giessen, Germany. karsten.muenstedt@gyn.med.uni-giessen.de

Very few human studies. More needs to be done.

Anecdotal Benefits:

• helps to regulate and balance hormones

helps lower blood lipids and cholesterol

• may prolong life span due to its antioxidant levels

• acts as a nutritive energy tonic

may help extend lifespan

• stimulates the immune system to fight viral and bacterial infection

• helps reduce cancer causing estrogens

may help those with Colitis or Irritable Bowel Syndrome

• Bone Loss

• bronchial asthma

• insomnia and sleep disorders

Helps to keep skin smooth, toned and elastic

• Promotes sexual vitality and rejuvenation

• Facilitates fertility and may help reverse impotence

• Has a bacteriocidal action on bacteria like staph

Helps to regenerate bone growth

• May help build tissue and muscles

• Supports wound healing

Protects the liver

• Increases physical strength

• Provides extra physiological support during pregnancy and menopause

Helps to reduce arthritic pain

• Stimulates memory and mental function•

may help with depression and anxiety

• Anti-anxiety

• may help boost metabolism

• May help with autoimmune issues such as rheumatism

Works Cited

www.Wikipedia.com

www.Mayoclinic.com

2005 Riviana Foods Inc.

www.hungryjack.com

Guyton, Arthur C., Hall, John E., **Textbook of Medical Physiology,** 11th Edition, Elsevier Saunders Inc, 2006

Nelson, David L., Cox, Michael M., **Lehninger Principles of Biochemistry,** 4th Edition, W.H. Freeman Co., 2004.

Abeles, Robert H., Frey, Perry A., Jencks, William P., **Biochemistry**, Jones and Bartlett Publishers, 1992.

Schweikhart, Melissa, Vanderbilt University, Nashville, **Tennessee Health Psychology,** 115A, December 10, 1996.

Schepdael, Van P. (1993), Effect of Ginseng G115 on the physical condition of triathletes, **Acta Therapeutica,** 19(4), 337-347.

Leung, Kar Wah, Wong, Alice Sze-Tsai, Pharmacology of Ginsenoside: A Literature Review, **Chinese Medicine,** 2010, 5:20.

Department of Obstetrics and Gynecology, Justus-Liebig University Hospital Giessen and Marburg, Giessen, Germany. karsten.muenstedt@gyn.med.uni-giessen.de

PMID: 18043624 [PubMed - indexed for MEDLINE]

PMID: 2391765 [PubMed - indexed for MEDLINE]

PMID: 14745176 [PubMed - indexed for MEDLINE]

PMID: 7556573 [PubMed - indexed for MEDLINE]

Experientia. 1995 Sep 29;51(9-10):927-35

PMID: 16951718 [PubMed - in process]

Department of Pharmacology and Toxicology, Faculty of Veterinary Medicine, University of Erciyes, Kayseri, Turkey.

Laboratory of Molecular Biology, Gifu Pharmaceutical University, 5-6-1, Mitahora-higashi, Gifu 502-8585, Japan. furukawa@gifu-pu.ac.jp.

PMID: 17213647 [PubMed - indexed for MEDLINE]

Department of Social and Environmental Medicine, Faculty of Medicine, University of the Ryukyus, Okinawa, Japan. Kaiissar@med.u-ryukyu.ac.jp

J Agric Food Chem. 2008 Nov 13.